Spin

Spin

*How the World
(and Almost Everything in It)
Turns*

Bill Gruber

McFarland & Company, Inc., Publishers
Jefferson, North Carolina

ISBN (print) 978-1-4766-9370-5
ISBN (ebook) 978-1-4766-5116-3

LIBRARY OF CONGRESS AND BRITISH LIBRARY
CATALOGUING DATA ARE AVAILABLE

Library of Congress Control Number 2023037039

Front cover images © 2023 Shutterstock

Printed in the United States of America

*McFarland & Company, Inc., Publishers
Box 611, Jefferson, North Carolina 28640
www.mcfarlandpub.com*

Contents

Introduction: Of Gyroscopes and Pitchers' Mounds   1

Chapter 1. Toy Story   7

Chapter 2. Around the House   28

Chapter 3. Amazing Grace   50

Interchapter I: A (Very) Short History of a Metaphor   67

Chapter 4. A Day at the Ballpark   73

Chapter 5. Big Wheels Keep on Turning   92

Chapter 6. Bullet Points   119

Interchapter II: Making Iron Come   137

Chapter 7. In the Sky   144

Chapter 8. A Descent into the Maelstrom   163

Chapter 9. Really, Really Big and Really, Really Small   187

Coda: James Clerk Maxwell and the Defenestration of Cats   211

Chapter Notes   213

Works Cited   221

Index   227

Introduction

Of Gyroscopes and Pitchers' Mounds

Everything is simple if you don't know a fucking thing about it.—KEVIN WILLIAMSON, *"Williamson's First Law"*

One Friday afternoon when I was still too young for school, my father brought home a strange-looking plaything called a gyroscope. It was round, about the size and heft of a baseball. But the toy was made of metal, so it was not some new kind of ball for sport. A disk attached to a spindle twirled freely inside a ringed, steel bracelet; a second ring connected at right angles to the first. Locked inside this skeletal sphere, the disk could be set spinning with the merest tweak of the fingers. It turned effortlessly, and, once set in motion, it seemed to take forever to stop rotating.

My dad showed me how to make the disk spin even faster. He threaded a short piece of twine through a small eye in the spindle, twirled the cord around the shaft in a tight, neat spiral, yanked swiftly and vigorously, and set the gyroscope down on a table. The toy throbbed and droned. It spun far longer than any top. We set it atop a wooden, tee-shaped pedestal where it spun upright for half a minute or more. I discovered I could also balance the spinning gyroscope on a taut piece of string or set it cockeyed on the tip of a sharpened pencil and carry it, securely perched, from one room to another. So long as the disk kept spinning, the gyroscope remained amazingly unperturbed no matter where or in what attitude I placed it.

Only after what seemed to be a cartoonish disregard for the laws of basic physics (as a five-year-old understood them, anyway), did the gyroscope slow, precess, and finally topple to rest on its side. Strangest of all was holding the spinning toy in my fist. I could feel it pulsing with energy. If I tried to turn it over, it resisted. The thing seemed to have a life of its own, weirdly non-mechanical. Whenever I moved it in one

1

direction or another, it seemed to come alive, like a kitten squirming in protest at being manhandled.[1]

That mysterious resistance I was feeling, I learned years later in Frank Hess's eleventh-grade physics class, was an effect long known to physicists as the conservation of angular momentum. Momentum of any sort is not an easy thing to describe, because unlike, say, mass or hardness, "momentum" is not an intrinsic quality or property of any object. Nothing stationary, no matter how big or small, has momentum. Objects possess it, whatever "it" is, only when they start to move. Simply defined, momentum (from Latin *movere*, "to move") is "mass in motion"; it's a fickle sort of property, in other words, that objects acquire whenever they're moving but lose as soon as they stop.

How do things move? That's the first great question of physics. When you slide the salad dressing to your dinner companion across the table, the cruet comes to a halt as soon as you stop pushing. This seems a reasonable way for things to behave. But suppose you pick the vessel up and toss it to somebody sitting on the opposite side of the room. It arcs swiftly through the air, then drops into a pair of waiting hands far beyond your reach. The thing somehow kept on moving even after you let go of it. That's a neat trick. How did you do it?

For children under the age of three, flinging toys and scraps of lunch is a common and (for them, anyway) hugely enjoyable pastime. Where toddlers find pleasure, however, ancient philosophers found mystery—and likely for similar reasons. One of the early, great riddles of mechanics was why objects set in motion kept on going even when nothing was touching them. How was it possible for stones to keep rolling, for spears to sail through space? Why didn't they just stop as soon as you let go of them? If you ponder that question, the answer seems neither intuitive nor easily deducible. It seems almost to involve some kind of magic—which is why the sight of Legos and mashed carrots flying down range still delights young children, even as similar sights once puzzled philosophers.

Explanations as to why stones and arrows continue in flight even when nothing seems to be propelling them had been proposed from ancient times. There was much doubt and disagreement. In one of his few utterly silly notions, Aristotle argued that projectiles were being thrust forward by the force of air continually closing around behind them. Other, cannier observers attributed the phenomenon to an "impetus" that had somehow been imparted to the missiles. They were right, of course. But in the late seventeenth century the older term "impetus" was abandoned in favor of "momentum," a word used (but almost certainly not invented) by the British mathematician John Wallis to describe the tendency of moving objects to continue in motion until some opposing force brought them to rest.

Intuitively, Wallis worked out the problem backward; he understood *momentum* to be whatever force it took to *stop* an object from moving. This became more or less the definition adapted by Isaac Newton in his *Principia*, a definition which is continued verbatim in modern high school physics texts. Something that moves in a straight line is said to possess "linear" momentum, which is defined as the product of its mass and velocity: *mv*. Even if only one of these two properties, mass or velocity, is great, their product obviously can be enormous. There's not much to choose, in other words, between being struck by a quarter-ounce bullet moving at a half mile a second and a twenty-ton municipal bus trundling down a city street. Both encounters would likely prove fatal.

"Angular" momentum is somewhat more complicated. It's defined as the mass of a spinning object multiplied not just by its linear velocity but also by the distance from the center of rotation to the center of mass. In other words, angular momentum is not simply the product of *two* factors *mv*, as for linear momentum, but *three: mvr*. Speeding bullets and city buses will continue straight in the direction they're headed until something acts to deflect or to stop them; their momentum, as physics texts inform students, is "conserved." A similar principle applies to angular momentum; it, too, is "conserved." But angular momentum, unlike linear momentum, is conserved in a way which seems not quite natural. Once in motion, any spinning object resists deflection from the axis around which it was first set rotating. As I long ago discovered fooling around with that toy gyroscope, such resistance can seem downright spooky. It's as if the spinning object somehow remembers its initial orientation in space and strives to maintain it.

Angular momentum—how strange, that poets and philosophers have not celebrated this transcendent reality. For this single principle— the conservation of angular momentum—makes possible an amazing range of everyday products and activities ranging from yo-yos and rifle bullets to salad spinners, electric drills, and flush toilets. Angular momentum is the reason a quarterback can throw an accurate forward pass to his downfield receiver; it's the reason bicycles wobble when getting under way but grow more stable at higher speeds, and it's one of the reasons, too, that airline pilots all over the world navigate their aircraft safely to their respective destinations. The conservation of angular momentum plays a role in natural phenomena like the spinning seeds of maple trees or maintenance of the Earth's magnetic shield, and it pertains to esoteric matters like the rate sensor units within the Hubble telescope or the conundrum known to nineteenth-century physicists as "the problem of the falling cat."

I wrote this book to describe and to celebrate this universe of spinning objects in layman's terms. Each chapter takes a few common things—salad spinners and rolling pins, an automotive wheel bearing, ballet dancers, a random, late afternoon "dust devil"—and attempts to explain how and why they do what they do. The book doesn't cover everything that spins—that would take an enormous book rather than merely a long one. To keep a nonfiction narrative from becoming ency-clopedic, I had to be selective, and readers are encouraged wherever they like to reflect on their own examples of stuff that spins.

If the narrative doesn't cover everything, neither does it always keep to a straight line. Basic physics, biographies, personal history, and the words and wisdom of experts all intertwine and overlap. My hope, in the words of the late novelist John Updike, is to give the mundane its beautiful due. It's easy to see grace in a perfect triple axel, to gaze in awe at the wheeling stars in the heavens, or to marvel at nature's ingenu-ity in "motorizing" certain kinds of bacteria with spinning flagella. It's less common to admire a washing machine drum, a wheel bearing, or a 55-grain bullet shot from an M16—though these things too, like stars and figure skaters, ought not to lack for wonderment.

Consider, for example, that .223 bullet: as it exits the muzzle, it is spinning, thanks to a set of spiraling, shallow grooves that have been cut into the inner surface of the rifle barrel, at a fantastic 3,000 revolu-tions per second, faster than almost anything else in the world, or, for that matter, in the universe. (The premier "natural" spin front-runner, a dying neutron star, the pulsar, PSR J1748–2446ad, is located in a cluster near the center of the Milky Way, where it turns round, in comparison with the M-16's bullet, a comparatively leisurely 716 times per second.) All these threads—skaters, bullets, wheels, rolling pins, baseballs, dust devils, planetary bodies, electrons, and many more, entwine in the fol-lowing story—or stories—of things that spin.

It started with baseball. For more than a decade I had been teach-ing a course in American Studies for Emory University called "Baseball and American Culture," and I was curious about the player who had first pitched from a mound and when and why he had done it. Baseball his-torians and encyclopedias yielded no conclusive answers, and explana-tions I found on various web sites were suspiciously thin. *Answers.yahoo* proposed that throwing downward off a mound increased the effects of gravity, thereby adding to the ball's velocity. Other writers claimed that an elevated position gave baseball pitchers the same advantage that enabled Hannibal's troops on the high ground above Lake Trasimene to rain destruction on Roman soldiers down on the shore. This theory at least sounded like common sense—I could easily picture a batter staring out and up at the 6'10" Randy Johnson, fearing he was about to be shelled by hilltop mortars.

Still, I wanted to know more about that strange little pile of dirt smack in the middle of the infield—its origin, like much of early baseball history, has been lost—and so I took my curiosity one afternoon to Emory's department of physics and walked the hallways in search of an open door. The man I barged in on that day, Stefan Boettcher, was gracious with his time and seemed genuinely interested to help me out. Boettcher was German-born and, once he learned the purpose of my visit, apologized up front for knowing nothing about the game of baseball. But as soon as I described the way a baseball pitcher throws from atop a mound, Boettcher cut instantly to the truth.

Pitchers standing on a mound derive no advantage whatsoever from gravity because of their superior elevation. This was nonsense, Boettcher said; the accelerative effects of gravity over a ten-inch drop would be all but non-existent. Also bogus was any comparison between pitching and occupying the high ground in warfare. Firing down from an elevated position on an enemy extends the effective range of bullets, arrows, or rocks, he explained, while missiles launched in retaliation upslope, owing to the effects of gravity, fall short. Obviously, there could be no such advantage in throwing a baseball from a mound.

Stepping down from a mound, Boettcher said, enables a baseball pitcher to take advantage of two quite different force vectors—translational (or linear) and rotational—each of which imparted additional velocity to a thrown ball. He drew two quick chalk sketches on the green board on his office wall. The first was a stick figure with one arm extended above the head in a throwing position, stepping forward along a horizontal line representing flat ground. The second drawing depicted a similar figure standing atop a small, gently sloping surface; this figure, when throwing, stepped both forward *and* down. "When a pitcher throws from level ground," Boettcher said, "the translational velocity of his forward stride is converted directly into the velocity his arm will impart to the ball. One linear force vector is simply added to the other: step speed plus arm speed equals ball velocity." Then Boettcher pointed to the second figure: "But if he is standing on top of a mound, when he throws, he steps not only forward but also down. Now his upper body must rotate around his front foot after it lands. This makes his head, shoulders, and arm move faster to keep up. This rotational velocity is added to arm speed and any translational velocity gained as the pitcher steps also forward."

Think of the way the tip of a blade on a wind turbine sweeps through the air at a faster speed—a *much* faster speed—than part of that same blade close to the hub. The tip of the blade moves faster because it must travel a greater distance in the same amount of time, just as the rim of a wheel spins faster than a spot near the axle, just as the outermost skater in the elegant, rotating lineup known as a "spoke" skims across the ice much faster than her companions closer to the center. Throwing from a mound, Boettcher said, would produce similar results.

The pitcher's throwing hand becomes in effect the tip of the whirling blade, and his stride foot, firmly planted on the downslope of the mound, corresponds to its hub. Boettcher did some quick calculations using an estimated eight-foot radius (which he took to represent a typical distance from a pitcher's extended, raised throwing hand to his stationary, leading foot) and a base hand speed at release of ninety miles an hour. As the doctor in physics sat at his desk and worked the numbers over, I could see that he was genuinely curious about what for him must have been a childishly simple problem. It was a pleasure to see a scientist fully engaged in discovering something about the way the world works. Finally, Boettcher turned to me, grinning. The additional velocity gained by pitching from off a mound, he said, as compared with flat ground, would be—his word—"significant."

Who knew? Gyroscopes, pitcher's mounds, Sufi dancers, wind turbines—from the biggest galaxies to the smallest subatomic particles, it was spinning all the way down. I bought a couple of college physics texts to see if I could learn again what I had been forgetting for half a century. I worked on questions involving rotational kinematics of propellers, figure skaters, bicyclists, gymnasts, divers, clothes dryers, ceiling fans, compact disks, drag racers, electric drills, merry-go-rounds, chainsaws, a quarterback's forward pass, roulette wheels, cassette tape recorders, and string trimmers. Mother Nature got into the act when I read about the rotation curves of galaxies, about cyclones and dust devils and the Coriolis Force, about maple tree seeds and species of bacteria that swim by using biological "motors" to spin hairy flagella.

Along the way, I even went back to that early baseball pitcher who first stood atop a mound. Using the same numbers as Professor Boettcher, I calculated that throwing from a 10-inch mound added about 3 mph to a pitcher's overall velocity as compared to throwing off the flats—significant, indeed! Learning about spin not only helped me to become more knowledgeable about curve balls and pitchers' mounds, but it also opened a route back to many of my boyhood interests, starting with that toy gyroscope. And it put me back in touch with a long-neglected self and a subject that was begging me to renew our acquaintance. This book is the story of that pursuit.

CHAPTER 1

Toy Story

Spinning Tops and Trundling Hoops

> Ajax caught up [a stone] and struck Hektor above the
> rim of his shield close to his neck; the blow made him
> spin round like a top [σβοῦρα, or whirligig] and reel in all
> directions [*Iliad,* trans. Robert Fagles].

Homer's description of warfare in Mycenaean Greece is likely as accurate as it is gruesome. Armed combat in the Bronze Age was at close quarters; to use longer-range weapons such as the bow was to display cowardice in the face of the enemy, a behavior shunned by real warriors. Men on both sides first hurled boasts, taunts, jeers; then came matchups, single-combat duels between champions. The warrior ethic was supreme, inviolable. Fighting in the Homer's *Iliad,* Hektor and Ajax hack and slash each other until their swords and spears shatter; then, lacking weapons, the pair bludgeon each other with heavy stones. Notice, though, how the vocabulary of children's play creeps into heroic diction. In this scene from the poem, battlefield violence engages Homer's imagination vividly and realistically only to be transformed ironically into metaphor. Ajax's blow is so brutal that it transforms a living man into a senseless, wobbling toy.

Spinning tops are among the oldest kinds of human playthings; invented independently in just about every clime and place throughout the world, the toys have been recovered from archaeological dig sites dating back more than six thousand years. Clay tops dating from 3500 BCE have been recovered from the ancient Sumerian city of Ur, and small, spinning toys fashioned from terra cotta or wood were discovered by archaeologists at dig sites in Troy (3000 BCE), Egypt (2000 BCE), China (1250 BCE), and Thebes (1250 BCE). A spin top is said to have been laid alongside King Tut in his golden tomb; for his use in the afterlife, the boy-king was supplied with jewelry, statues, weapons, a chariot, a bed in the form of a lion, changes of clothes, and a carved, egg-shaped toy for spinning. A spin top was one of seven toys (along with mirror, dice, ball, apple, hoop, and tuft of hair) given to the infant Greek god Dionysus, patron deity of wine, vegetation, festivity, and madness.

The dance of a spinning top evokes wonderment and mystery. The toy is full of incongruities, contradictions—fun for children, captivating for adults. In the *Republic*, Plato writes of tops which, "when they spin round with their pegs fixed on the same spot, are at rest and in motion at the same time" The poet Virgil in *Aeneid* compares the tragic Amata, the mother of Lavinia, to a whirling boxwood that "wanders aimless ... made a living thing." Not only philosophers and poets have been drawn to contemplate the spinning, humming playthings. Early physicists wrote entire books about tops; graduate students in contemporary physics departments still publish analytic papers about them.[1]

The famed Swiss mathematician Leonhard Euler spent twenty years developing equations to explain the spinning motions of rigid bodies about a central axis, bringing into his calculations everything from toy tops to heavenly planets. There's something both elemental and universal about the toy; even the fundamental particles of the universe act as if they were tiny, spinning tops. A photograph taken in the early twentieth century captures two famous quantum physicists, Niels Bohr and Wolfgang Pauli, peering down in thrall at a toy top spinning by their feet!

You cannot stand a top on its slender foot even for a moment. Motionless, severely out of balance, it's unstable as a one-legged stool. But twirl the same top between your fingers so that it's spinning when you set it down, and it will remain upright for a very long time. Spinning, it will wander, knock into things, even circle back to where it started, but it won't fall over until just before it stops whirling round. Like a gyroscope, a top doesn't like to be pushed around; its spin axis, as Plato noted, remains unmoving. A spinning top stands upright on its base because the angular momentum imparted initially by the snap of index finger and thumb is conserved along its axis of rotation.

Angular momentum is a vector quantity; it has a specific direction, and it takes another, independent force to redirect it. When a top is spinning upright, its angular momentum is also pointing straight up; technically speaking, its axis of symmetry and its axis of rotation coincide. Flick the toy with your finger: so long as it's spinning rapidly it won't fall down. Tip it carefully sideways: the top will modify its spin axis to a slanting position and keep on twirling, obediently conserving its angular momentum along this new vector. In a world without friction, a spinning top would defy gravity indefinitely.

Sooner or later, though, gravity pulling on a spinning top will lever it a little sideways. Once that happens, the axis of symmetry and the axis of rotation no longer align. If the top were stationary, at this point it would fall right down. While it spins, however, it won't keel over just yet. Rather, it will lean slightly, and simultaneously its upper end will start to revolve slowly around an imaginary, vertical line extending up from its supporting foot. This slow, circling movement around a virtual axis is called precession, and the speed at which the top precesses is inversely proportional to the speed at which it is still spinning. As friction gradually causes the top to spin slower and slower, the toy simultaneously circles faster and faster around the vertical axis in what looks like a final, willed attempt to remain upright.

The tug of war between the forces of gravity and angular momentum lasts for some time. Sooner or later, however, the pull of gravity is sufficient to lever it rudely down on its side where it comes to rest with a few final, skittering turns. More technically: the top spins upright until friction between its pointed base and the surface on which it spins slows it to the point where the torque exerted by gravity on the object's center of mass overmatches its quantity of angular momentum and pulls it over and down.

Spin tops are not just for play. Perhaps the most widely known among ancient tops is the four-sided dreidel, created (as legend has it) by Jews to learn the Torah in secret after public use of the sacred text was forbidden. A common version of the dreidel story holds that when the study of the Torah was outlawed at some time during the second century BCE, Jews hid their expressions of faith out in the open, so to speak, by making it look like they were not praying but gambling—a public pastime the Romans considered normal, even healthy. Jews took a common spin top (then known as a teetotum) used widely by Roman gamers and gamblers, changing the lettering on the sides to serve as mnemonic tools for learning the Torah.

Among the various combinations of original letters on the original top were A, D, N, and T. For gamblers, these meant, respectively: *aufer* (Latin for "take from the pot"); *depone* ("put into the pot"); *nihil* ("nothing"); and *totum* ("take all"). Players took turns spinning the dreidel in much the same way they alternated throws of dice. The modern dreidel associated with the celebration of Hanukkah, on the other hand, is imprinted on each face with a different Hebraic letter—*nun*, *gimel*, *hey*, and *shin*. Together these letters are said to encode the message "Nes gadol hayah sham," or "a great miracle happened there," supposedly a reference to the recapture of the Temple in Jerusalem in 165 BCE (Rosenberg 2014).

Spinning tops have appealed to artists as well as to children, philosophers, worshippers, and gamblers. In his encyclopedic depiction of contemporary children's games, Pieter Breughel the Elder paints several boys inside a portico, each bent intently over his own spinning top. The boys hold short sticks with a string attached to one end. To set these "whip tops" whirling, a boy would have wound the string around the top, steadied it upright with one finger, then yanked the stick quickly to unwind the cord and set the toy spinning and skittering across the stone floor. Such whip tops were popular with children in Europe and America for hundreds of years, well into the twentieth century. Don Olney, who used to run a toy store in Rochester, New York, has been making and collecting these and many other kinds of tops for years; he's got tops from different cultures and times, tops that play music, tops that draw swirling, abstract pictures, magnetized tops that will spin hanging upside down from a paper clip. Olney sees tops in much the same way as did Plato. "Tops do something our brains tell us they shouldn't do," he says. "They're definitely metaphysical in some way" (Kastor 2021).

Industrial furniture designers Bernice ("Ray") and Charles Eames made more than eighty documentary or "exploratory" films in which they worked out various aesthetic problems connected with different objects ranging from chairs and toy trains to cemeteries and the textiles of India. In *Tops*, one of their so-called toy-films, the couple celebrate the ancient art of top-making and top-spinning. Their homage contains neither dialogue nor narrative voiceovers. None is needed; the tops speak eloquently for themselves, seven minutes of spinning shapes and young kids' enthralled faces. There are tops with slender cords twisted round their spindles, brightly colored tops spinning on the surfaces of tables or floors, some leaving tracks across the sand; others tickle the palms of children's small, upturned hands. There are shots of the undersides of tops, filmed as they spin across glass tables; still other film sequences show tops as they are reflected in mirrors or in the shining eyes of beholders.

All the shots in the Eames' film are done close-up. There's no dialogue, no interaction between the different children, no narrative other than that provided by the changing focus of the camera, accompanied by a musical score composed by Elmer Bernstein. Slim contemporary designs spin next to crude, ungainly shapes; there are tops made of ceramic, epoxy-and-resin composites, various metals, historic toys carved from pine, beech, maple. The tops shimmer and sparkle, bundles of color, physics, whimsy. They blend things counter, spare, strange.

A slender, wooden twirler traces lines precise as if drawn for a textbook on graphic design; another, carved in the shape of a dancer pirouettes effortlessly on a single, pointed foot. A top carved in the shape of a blooming flower spins with petals open, its multiple colors smeared to a blur. Still another top is blob-shaped, blatantly commercial; on it spins an advertisement for Cracker Jack candied popcorn. Yet it too whirls with a ballerina's grace; angular momentum doesn't discriminate. According to the Eames' website, the film has proved useful not just to celebrate the beauty and art of this ancient, spinning toy but to showcase basic principles of motion; for years it was used by MIT professor Philip Morrison who played it for students learning rotational kinematics.

Spin tops have lost their former place in the hierarchy of children's pastimes, but facsimiles have made serious inroads into the world of adult chic. As I write, the website foreverspin.com offers for sale numerous miniature tops machined from eighteen different metals. One can select tops machined from aluminum, copper, and stainless steel (about $40 each) on up to more exotic substances such as black zirconium ($94), silver ($164), and tungsten (advertised as "the heavy one," weighing in at just over two ounces and costing $195). Foreverspin also sells "starter packs" of five different tops as well as elegant "museum sets" of sixteen different tops; each such set comes with a metal display dock to store the tops when they are not in use, also a circular, metal pad on which to set the toys twirling. These pads are micro-polished to minimize friction and so to lengthen spin duration.

There's a wide spectrum of performance among the different kinds of tops. Since the moment of inertia of a rotating object increases proportionally with its mass, a top made of tungsten will carry more than twice the rotational energy as one machined from stainless steel and nearly seven times as much as a lightweight made of aluminum. The heavier the top, of course, the harder it will be to get it moving, but once it's spinning it will take longer to slow down. Sometimes much longer. A recent YouTube video shows one such Foreverspin top, set in motion by a Dremel rotary tool, spinning merrily (if somewhat eerily) on its pad for six mesmerizingly long minutes.

Almost as old as the top is another familiar kind of spinning toy, the hoop. The young Dionysus wasn't alone in playing with both. Children in pharaonic Egypt twisted grape vines into circles and rolled them across the sand, spun them round their waists. In South America, similar hoops were made from the sugar cane plant; fashioned from strips of wood and metal, hoops exploded into a craze in fourteenth-century Britain. When Christian missionaries disembarked on the Pacific islands of Hawai'i, they saw dancers twirling large, slender rings around their waists; the visitors called the rings "hula hoops" after the native word for the swaying, Polynesian dance.

A century later American teenagers were performing the same dance moves with colored, plastic hoops. In 1958, a hula hoop could be had for a buck ninety-eight. Americans bought twenty million of them in the first six months they were on the market; 100 million were sold worldwide. "No sensation has ever swept the country like the Hula Hoop," says the journalist Richard Johnson, author of *American Fads*; the hoop "remains the one standard against which all national crazes are measured" (Stevenson 2020).

Romans inherited hoop-rolling, as they inherited so many things Grecian. In his *Art of Poetry*, Horace listed hoop, ball, and quoit as expressly manly sports requiring both training and skill; the poet Ovid, writing from exile, dreamed nostalgically of children playing in the streets of the beloved city he saw now only in his imagination: "now flies the ball, now rolls the whirling hoop" (*Trista III*, elegy 12). A mosaic from the sixth century CE shows two boys playing with spoked hoops; this artwork once decorated the Imperial Palace of Constantinople. On a wall no one currently knows where—perhaps in Germany, perhaps in South America—there hangs a portrait of the young Pedro Alfonso, Prince Imperial of Brazil. The work of the nineteenth-century German artist Johann Moritz Rugendas, the boy prince poses with hoop and rolling stick for a portrait commissioned by proud parents only months before his sudden death from epilepsy at 2 years, 108 days (Alfonso I, commons.wikimedia.org).

"The hoop was also a significant image in 17th-century Dutch art," says Milly Heyd, where it took on "both positive and negative moral connotations"—positive for its associations with children's carefree play, negative, according to Heyd, "because trundling of the hoop has a Sisyphean quality ... of which we are usually unaware" (Heyd 97).The image of child and hoop became the focus of the Greek expatriate Giorgio de Chirico who painted "The Mystery and Melancholy of a Street." The painting shows a piazza in yellow, evening light, some arched, blackened doorways, and the shadow of a statue, perhaps military. A girl runs along the empty street, rolling a hoop, hair and skirt flying behind her; she and her hoop are the only moving things amidst the stillness.

Greek and Roman play hoops were sometimes purpose-built out of metal (typically bronze, iron, or copper), but the Roman poet Martial (first century CE) indicates that the thin, hammered metal tires used on cart wheels could also be repurposed as children's playthings. The sport of hoop rolling was popular with young Eurasian youths who trundled them on the frozen Danube River. Beer-drinking cultures needed millions of thin, circular metal "stays" to hold wooden barrel staves in place, and boys and girls almost universally adopted these thin, round strips for use as toy hoops.

Later, when the supply of barrel stays ran low, there were always bicycles. At the turn of the nineteenth century in America, by which time bicycles had become ubiquitous, a repurposed wheel rim could be turned into an excellent, readymade hoop. The vacated rim was rolled along the ground by stroking it with a short stick; once moving, angular momentum generated by the spinning hoop kept the plaything headed in a

The mythological Ganymede ("most beautiful of mortals") trundling a hoop; the youth is depicted on a vase created by the "Berlin Painter" (an Attic Greek vase-painter of the fifth century BCE) (Louvre Museum, Campana Collection).

straight line. Repeated, gentle strokes kept one's hoop moving for long distances at whatever pace one desired, and skilled "trundlers" could make their hoops change directions or navigate even around street corners. Hoop-rolling was popular with children and adolescents well into the twentieth century until increased automobile traffic made it difficult (and highly dangerous) to roll the toys along city streets.

The sport of hoop-rolling, like other children's games, has been frequently depicted by artists. Hoop-rolling is central to Bruegel's "Children's Games" (1560); in "Children's Hoops" (linocut, 1936), the twentieth-century Australian artist Ethel Spowers shows a group of children pursuing their hoops over the crest of a hill and out of sight. Hoop rolling seems to have been most popular among pre-adolescent boys, but not exclusively so. "Girl with a Hoop" (1885), a well-known impressionist painting by Pierre-Auguste Renoir, depicts nine-year-old Marie Goujon; she stands in a garden holding a large, wooden hoop in her left hand, a short "trundle stick" for propulsion in her right. In another work from the same period, "The Umbrellas," the artist seems infatuated with roundness, with the delicate, curving rims of umbrellas, bonnets, picnic baskets. In the foreground and completing the scene, slightly off to the right, a young girl clutches her toy hoop and stick, staring out at the viewer.

The hoop may have been the first toy ever to bend gender boundaries. In ancient Sparta, girls spun hoops down the streets as part of their education; at the same time, 800 stadia to the north, grown men rolled them across the agora in Athens because the physician Hippocrates recommended playing with the hoop as a cure for weakness. In one of the earliest images of hoop-rolling, the so-called "Berlin painter" of the fifth century BCE depicts the young Trojan hero Ganymede (the "most beautiful" of mortals). As Ganymede runs, he somehow balances a rooster on his left arm while trundling a hoop with his right.

Two thousand years later, Victorian graduate students at Cambridge University trundled hoops to relax after a day of head-spinning lectures; during the same era, hoop-rolling (along with swinging jump ropes and dumbbells) was a standard part of young girls' physical education. It's still part of the education of young women at Wellesley College in Massachusetts, at least in a manner of speaking. Every May Day, graduating seniors assemble on campus at Tupelo Lane and stage a hoop-rolling race down the hill toward Lake Waban, competing for the winner's prize of being tossed into the water by the losers. Once upon a time the so-called winner was predicted to be the first in her class to marry. In more egalitarian times, the hoop-rolling crown goes to the person who first achieves "success—however she defines it" (wellesley. edu, 2012).

The physics of hoop-rolling are simple. A hoop is basically a disk with most of the inner material removed. The rotational inertia of such an object depends not only on its overall mass but on *where* that mass is concentrated with respect to its spin axis. The farther the mass is located away from the center of rotation, the greater the moment of inertia; for example, as it rolls down the street, a thin iron or steel hoop weighing about a pound will have about twice the rotational inertia of a solid disk of the same diameter and weight. (The equations for calculating the rotational inertias of hoop and solid disk, respectively, are $I=MR^2$ and $I = \frac{1}{2}MR^2$.)

The displaced location of the weight will make the hoop slightly more tippy than the disk, and its greater moment of inertia means the hoop will be somewhat harder than a disk to start rolling in a straight path. In contrast to the solid wheel, however, as soon as the hoop is moving it will become *much* more stable. Once rolling, the greater the hoop's diameter, the greater its rotational inertia and the more it will tend to resist being bumped off course. A relatively large hoop (the one Ganymede rolls on the vase in the Louvre, for example, extends from his foot almost to his armpit) will tend to keep rolling in the same direction for a surprisingly long time, needing only the occasional nudge with a stick (palms and fingertips tend to be too grabby) to coax it along. Historically, the best-performing hoops tended to be made of thin metal strips several feet in diameter—happily, for many dozens of generations of children, wagon wheel tires, coopers' stays, and bicycle rims were almost always abundant.

Chillin' Out

Not included in the babe Dionysus' toy chest—but likely as old as the rest of his playthings—was a small disk-on-a-string called a yo-yo. Like hoops, tops, and gyroscopes, yo-yos mix rotational kinematics with the stability provided by angular momentum. Hold the toy in your hand, one end of the string attached to a finger. Clutch a stone about the same weight in your other hand, then let both stone and yo-yo go. Gravity will accelerate the toy down, just as with the stone. But the yoyo will drop a little slower than the stone because some of the energy associated with falling will pull on the string and start the toy rotating around its axle. The farther the yo-yo falls, the faster it spins. When it reaches the end of its string, of course, it can't go any farther. But it will still be spinning quite fast, carrying a lot of rotational inertia that it acquired as it fell downward, and that energy will be sufficient to re-wind the toy all the way back up to your waiting hand, where the cycle begins anew.

Or almost all the way back up: frictional losses will bleed off some inertia with each successive climb back up. Each return up the string will be a little lower than the previous one, until eventually the toy stops spinning and dangles motionless at the bottom. But you can keep the thing bobbing up and down forever if you replace the lost inertia by giving the yoyo a quick flick of the wrist each time you send it spinning downward.

Yo-yos, like tops and hoops, come with a hoary pedigree; the toy itself is almost certainly ancient, but not its name, which was coined in 1928 by Pedro Flores, a Filipino immigrant to the United States who began manufacturing the toy in Santa Barbara. Flores may have adapted the name yo-yo from an Austronesian term *yóyo*. Before Flores began selling yo-yos similar spinning toys were called, among other names, *bandalore* in France, *quiz* in England, and *chucki* in eighteenth-century Europe (Goto-Jones 2015). A Victorian publication *All the Year Round* (edited by none other than Charles Dickens) includes a story entitled "Quite a Lost Art" in which characters play with small, spinning toy called "the ivory bandalore." In that story, a hermit steals an ivory bandalore from a visiting prince; the toy had been made as a gift to the king of Cordova to amuse his children.

Convincing evidence of yo-yo's ancestry comes from the visual arts. Berlin Museum's *Antikensammlung* [Antiquities Collection] includes an Attic *kylix*, or shallow household bowl, from the fifth century BC showing a boy dangling a small, round object from a string. On display in the Louvre hangs a painting from the late eighteenth century done by the artist Elizabeth-Louise Vigée-Le Brun. Her portrait shows a golden-haired boy, believed to be the young Louis XVII, playing with his *yoyo emigrette*. Adults throughout history seem to have played with the toys almost as much as did children, using them to fill empty hours much the same way a bored adolescent might pass the time today playing Hateful Boyfriend on a smart phone.

A small, eighteenth-century watercolor from the Brooklyn Museum, "Lady with a Yo-yo, Northern India," for example, shows a woman amusing herself by standing on a stool and swinging a round, string toy; the painting is said to belong to an extensive genre of such works depicting lonely women who had to find ways to pass the time while their husbands and lovers were absent.

To attempt to mend a broken heart by playing with a yo-yo may sound silly, but such behavior is consistent with everyday experience, as the therapeutic use of yo-yos has been well-documented, says aficionado Chris Goto-Jones, a professor of philosophy at the University of Victoria. Writing on yo-yos in *The Atlantic*, he maintains that it "gives a particular kind of thrill to feel in control of such a dynamic, energetic, and animated object" (Goto-Jones 2015).

Another yo-yo enthusiast, Charlie Yo-yo (yes, that's his real name), agrees. Mr. Yo-yo makes yo-yos at the Omega Factory in Fort Lauderdale. "We're teaching kids the Zen of the yo-yo," he says. "You can't let the thing just drop; you've got to bring it back up. That's what yo-yo-ing is all about, finishing the task. It's great because it's so relaxing. It's absolutely the most stress-free sport" (Yo-yo 1997). Charlie Yo-yo might be onto something. In revolutionary France, the toy was believed to be so effective in relieving stress that the nobility were said to have played with *yoyo emigrettes* during their ride in tumbrils to the guillotine. Or perhaps the nobles had picked up the idea from watching Beaumarchais's *The Marriage of Figaro* (1784), in which the nervous hero claims his yo-yo to be "a noble top, which dispels the fatigue of thinking" (Oliver n.d.).

A small, eastern Pennsylvania town in mid-century America

offered no such existential threats to childhood innocence. Still, all through fifth and sixth grades my classmates and I were hooked on yo-yos. Our play with the colorful, spinning toys bordered on obsessiveness, especially during those late winter months when melting snow berms and drizzling rains made both the schoolyard and nearby athletic playing fields too sodden to use. There were in those years no downloadable instruction manuals or YouTube videos where we could learn about the various maneuvers that might be performed with yo-yos, much less the skills necessary to perform them. Technique and knowledge of what was possible with our toys came only through watching older boys perform "loop the loop" or "rock the cradle," then rehearsing the tricks ourselves until we had achieved some vague semblance of the ideal.

We could select from four distinctly different models of yo-yos ranging in price from about a quarter (a significant sum when candy bars sold for a nickel) up to a dollar. Aesthetics and quality of manufacture naturally were on a finely graded scale that improved in lockstep with rising cost. The cheapest yo-yos were pudgy, wooden disks, painted in solid colors—one side a dull barn red, the other, basic black. These clumsy toys were useful only for learning the bobbing arm motion and quick wrist snap necessary to impart enough spin and downward motion to cause a yo-yo to drop and roll right back up its extended string. Only linear, up-and-down motion was possible. The string was stapled at one end to the spindle; once the cord hit its limit, angular momentum took over and rolled the yo-yo right back up.

The more money we spent on a yo-yo, the greater the possibilities for amusement (not to mention bragging rights) that opened. Shiny colors and flashy graphics were important, of course, but our purchases were not just a matter of aesthetics. One step up the food chain was a yo-yo with a different method of attaching string to spindle. Unlike a beginner's cord, which had its end fixed to the axle, this string was doubled, twisted, and one end looped around the axle rather than permanently stapled to it. This made it possible for the disk to continue to spin freely while it dangled at full string extension—to "sleep," as the trick was called—and *that* made it possible to take advantage in all sorts of ways of the conservation of angular momentum.

One could, for example, lower the sleeping yo-yo gently until it just kissed the floor, then "walk the dog" forward several feet before a quick wrist snap re-established friction between string and axle and brought that puppy right back up to the top of its leash. Best of all were Duncan "tournament" models. These were glossy, thickly enameled toys painted with exotic maroon and electric blue enamels and decorated with four iridescent diamonds inlaid across the diameter.

Yo-yo-ing remains a sport for the young. The average age of competitors in national and international yo-yo championships is between 11 and 21 years old (Pogue 2018). More than forty years after I had cast aside my own yo-yos for other pursuits, a similar yo-yo mania took hold of my son and his middle school circle of friends. By the first decade of the twenty-first century, the toys had become both expensive and lavish; one of my son's models featured multicolored stroboscopic lights and an eerie, electronic whistle. But the maneuvers themselves were timeless and might have been performed by George McFly in *Back to the Future*: "Rock the Baby," "Sleeper," and "Around the World" were some I remembered from my own barbarian days.

If spinning toys really did help *la noblesse* relax on the way to meet *Madame la Guillotine*, imagine how they might brighten the outlook, say, of squirrely children on a rainy afternoon or a systems analyst at a three-day corporate retreat. Such were the original motives for the creation of fidget spinners. These tiny, hand-held mechanisms were invented by Catherine Hettinger in the early 1990s. At the time Hettinger was living in Winter Park, Florida, and suffering from myasthenia gravis; unable to pick up her young daughter's toys "or play with her much at all," as Hettinger says, "I started throwing things together with newspaper and tape, then other stuff. We kind of co-invented it—she could spin it and I could spin it, and that's how it was designed" (Luscombe 2017).

Hettinger took out a patent on a tiny, spinning toy (US5591062A) with a "center dome structure and a skirt ... designed to be spun on the finger" but was forced to relinquish it in 2005 because she could not afford the $400 renewal fee. Ten years later, when manufacturers and toy stores were hauling in boatloads of profits from global sales of a look-alike, second-generation "fidget spinner," the toy's cash-strapped creator was downsizing her life to a cheap condo and wondering how to pay for (her words) "a car that truly works." But Hettinger doesn't feel bitter about the riches that might have been hers; she's just pleased to know that her invention has proved to be much more than a simple toy. "I know a special needs teacher who used it with autistic kids," she says, "and it really helped to calm them down" (Luscombe 2017).

The most recent iteration of the fidget spinner rose from the daydreams of Scott McCoskery, an IT engineer from Seattle. McCoskery has said he created the toy—at first he called it a "torqbar"—to help him overcome boredom during conference calls and in business meetings that he didn't want or need to attend. "During those times," McCoskery said in an interview with NPR, "I often found myself clicking a pen [or] opening and closing a knife, just to get through the meetings" (Malone 2017).

Fidget spinners may have been created by adults, for adults, to help soothe the stresses of modern corporate life. But in what opens a sobering view into the world we have made for children, the biggest market for McCoskery's toys turned out to be not bored office workers but school kids. By 2017, pre-teens and teenagers were buying and using fidget spinners by the tens of millions. The whirring, purring gadgets soon became ubiquitous nationwide in hallways and classrooms; they were most often brought out during instructional hours. One would have hoped that someone, somewhere might have put two and two together—the toy was, after all, created to help its users resist ennui. One would have hoped that the people in charge of children's education would have first tried to understand why contemporary instructional methods were driving kids to obsess over their fidget spinners. Alas, most schools simply banned them.

Of Pie Tins and Fly Swatters

Older than fidget spinners—but much younger than tops or hoops—are the spinning, flying-saucer-shaped toys called Frisbees, named so after William Frisbie who founded a pie-making company in Bridgeport, Connecticut, shortly after the close of the Civil War. Then, as now, the small state of Connecticut was clustered with colleges, and hungry students from nearby schools would first chow down on Frisbie's pies and then spin the empty tins back and forth, shouting "Frisbie!" as they let the lightweight discs fly. The temptation to fling empty pie tins through space seems inborn.

Even as the first plastic toy disks were being manufactured in 1957 by the Wham-O corporation, I was hurling my grandmother's surplus pie tins skyward, pretending they were miniature alien spacecraft come to roost in the back yard. It seemed a natural thing to do with the empty tins, and maybe it was; the psychiatrist Dr. Stancil Johnson believes that people the world over naturally want to throw empty pie tins because the activity marries humans' greatest tool—the hand—with their loftiest dream—to fly (Latson 2015).

I had a conversation about Frisbees recently with Jim Herrick. Herrick is kind of a big deal in the world of Frisbees; he's been playing with the toys for almost fifty years, and along the way he's won several world championships (1982, 1983, and 1988) in one of the iterations of the sport called "ultimate Frisbee," or simply "Ultimate." Herrick came to Frisbee by way of basketball. He played hoops through high school, made the team at Cornell University as a freshman walk-on, even played a couple of minutes in a win that season against then perennial Big East

Conference powerhouse Syracuse. But he was cut from the team the next year, and as he made his lonely way back to the dorms that afternoon, Jim Herrick took stock of his vanished basketball prospects. "I was thinking," he said, "this is all I've done for ten years. What do I do next?"

He stumbled across the answer to his question within minutes when he spotted a handful of students dashing back and forth on one of Cornell's athletic fields, playing something that looked a little like soccer, a little like football or rugby, maybe also a little like lacrosse. Except that the players weren't tossing or kicking any sort of ball; instead, they were flinging a colored, plastic disk.

"I jumped into the game immediately," Herrick said. "And that just led to it." He paused while he contemplated the strange paths a life can take. "You know, in a lot of ways it really defined where I ended up living, who I married. The sport was just starting, and all these guys went to colleges in the east, and they started teams. We had a good team. Bill Nye the Science Guy was on my team."

When people new to the sport first attempt to throw a Frisbee they imitate the *Discobolus* of Myron, torquing their upper body and arm and flinging the disk up and out with a twisting, backhand motion. But modern players most often throw a Frisbee with a delivery like a side-arm baseball pitcher. It's called an overhand wrist flick, a throwing technique Herrick and others pioneered in the early days of the sport, and it enables players to impart enormous spin rates to the Frisbee as it leaves the hand. In fact, Frisbee is all about the spin, Herrick said. "They call it Z's. You've got to hit a lot of Z's. If you have a lot of Z's the disk is spinning really fast, and you can do more tricks."

The athletes who play Frisbee may not sign multi-year, half-billion-dollar contracts, but neither are they a handful of amateurs living in their vans. "Now there's a half dozen people making a million bucks a year," Herrick said. "There's professional leagues, college championships, a big international presence." The sport also is growing in high schools, he says. "They [the high schools] realize, oh, wow, the football team is costing us 400K. And Frisbee gives us the same benefits for a tenth the price and without the concussions."

A Frisbee's aerodynamics are not complicated, but they're interesting. It turns out that those nineteenth-century college kids were onto something. An upside-down pie tin is an efficient flying machine. When Nature designs creatures for flight, she combines light weight in relation to broad, horizontal surfaces. An empty pie tin checks both boxes. Next comes aerodynamic lift, and here, too, a pie tin fits the bill. Slice across its hollow interior and sloping sides and you're looking at a shape remarkably like the cross-section of a wing. This profile—shallow curved surface on top, under-cambered below—provides the buoyancy necessary for extended, gliding flight.

When a pie tin is thrown at a slight positive angle, air moves over the convex, upper side (which is its bottom) slightly faster than it moves along underneath. According to Bernoulli's rule for fluid mechanics, the different velocities result in a pressure differential between top and bottom (greater below and lower above), causing the disk to rise. Finally, when a pie tin is tossed into the air with a quick, hard sideways flick of the wrist, it acquires a significant quantity of angular momentum; this keeps it flying on a stable trajectory.

The rolled, metal edge of the pie tin, like the thickened rim of the modern, plastic toy, biases weight further away from the spin axis, slightly increasing the pan's rotational inertia. You want to launch pie tins and Frisbees spinning at around 500 rpm. The spin is key; because a Frisbee's center of lift and its center of gravity are in different places, a significant quantity of angular momentum is required to overcome these imbalanced forces. If the disc isn't stabilized by spinning rapidly, those contrary torques, Bernoulli pushing up and Newton pulling down, will start the Frisbee seesawing and tumbling end over end.

If you cut out the center of a Frisbee leaving only an empty circle of plastic, the remaining mass will be concentrated even further away from the center of rotation, and the increased rotational inertia (according to the same physics as with rolling hoops) will help your toy soar stably much farther down range. Distances of four hundred feet or more are easily achieved even by amateurs with these spinning, gliding rings, and, in 2003, one such flying machine called the Aerobie Flying Disc spun elegantly through the air for a quarter of a mile (400 meters), more than a lineup of four football fields. This was—and still is—farther than anything else in the world, whether rocks, spears, baseballs, boomerangs, or nunchucks, had ever been thrown by hand (Lissaman and Hubbard 2010).

Dig deeper in the toy box. Consider Skittles, a miniaturized version of an old European lawn game from which the modern sport of bowling is derived. As with the game of nine-pins played by Rip Van Winkle and the ghosts of Henry Hudson and his crew, early skittles took place outdoors, often on grassy fields near taverns. But the sport soon migrated indoors where participants could bowl on purpose-built alleys in pubs and hotels. In time, smaller and still smaller versions were developed that could be played on the parlor floor or even on dining tables.

I first played the table-top game when I visited the childhood home of my wife one Christmas. After the presents had been opened and sufficiently admired, I was informed by her family that it was time to play the traditional holiday round of something called "skittles." Out from a closet came a shallow wooden box about three feet long and a foot and a half wide. The box was divided into eight or ten different compartments with small pass-throughs between them. Instead of a ball, tabletop skittles used a small, spinning top to knock down wooden pins set upright in all the different parts of the box. The pass-throughs permitted the top to migrate from one space to another, blundering in and out of different rooms like a drunken partygoer knocking down the furniture. The carnage lasted until the spin top ran out of pins, or angular momentum, or both. The toy I played with had been made in 1940s America, but you can buy this old-timey plaything even today; the contemporary version is manufactured, appropriately enough, by a company named "Carrom" who pitch it as a "tornado in a box."

In search of more spinning playthings, I took a quick skim through a catalogue from Hammacher-Schlemmer, a long-running mail-order company who proclaim a history of "guaranteeing the best, the only, and the unexpected for 171 years." Hammacher-Schlemmer advertises an eye-catching display of household gadgets that are almost offensively superfluous to the examined life. You would never think of such products on your own, but somehow, they all seem highly desirable once you see them. Perhaps it is because these esoteric toys are so emphatically unnecessary that they appeal to the oligarch latent in most of us.

Among the items for sale in the spring of 2019, Hammacher-Schlemmer advertised a spinning Easter egg decorator, a Mickey Mouse watch for women with a rotating dial depicting historic images of Disney's famous cartoon character, an end table that revolves to provide easy access to magazines and books on all four sides, and a battery-powered fly deterrent that shoos away insects by means of two slow, contra-rotating propellers. This device—sadly, it seems no longer to be available—was said to be practical as well as "eco-friendly."

Twin, ultra-lightweight, 15-inch blades spun slowly atop a cone-shaped base, causing a gentle, turbulent breeze that kept nosy bugs from food and drinks without harming them. Most people who purchased the eco-friendly fly deterrent seemed happy with it—it was "noiseless, harmless, unobtrusive, and interests everyone from curious, fingering toddlers to dotty adults," as one reviewer put it (Hammacher Schlemmer 2020).

There's something hypnotic about things that spin in super-slow motion. They provide a kind of cognitive fascination, a psychological "binding" that is accomplished exclusively through the portal of the eyes. Perhaps it's because their nightmare slowness makes us acutely conscious of existential duration, either overtly, as with the circling of a second hand around a clock face, or surreptitiously, in the case of the relentless and distinctly melancholy turnings of the blades on a large ceiling fan like the one in Rick's Café Américain in the film *Casablanca.*

Objects that spin slowly are used also to induce states of relaxation or calm. The Canadian "mocumentary" television series, "Trailer Park Boys," made use of this effect in a scene (2002) where Mike Smith, who plays a young, mildly autistic man called "Bubbles," mourns the loss of his 1961 Electrobubble, a Christmas gift long ago from his parents. Before it was destroyed in a house fire, Bubbles' gadget worked by blowing a stream of air against a wheel of small rings that were dipped one-by-one into a tray of soap solution as the wheel spun slowly round.

Though Smith's Electrobubble is entirely fictional, there is no shortage of similar, mechanical bubble-making apparatuses on the market, vintage or contemporary. Most work on the same design, slowly rotating a circle of small rings so that some are always being dipped in a soapy liquid at the same time a fan blows air gently against others. "This here's my pride and joy," actor Smith said of his beloved, bubbling device. "Works like a charm too, every kitty I've ever met loves these things."

Wheels of Fortune

Sigmund Freud wrote of a game played by his young grandson in which the boy repeatedly threw a wad of cotton out of his crib so that his mother would have to return it. Beneath what looked like repetitive, mindless play, Freud saw a subtle, calculated attempt by the young boy to cope with the appearance and disappearance of his beloved mother, over which he had no power, by manipulating the comings and goings of a plaything, whose movements he could direct at will. Throughout history, games of chance involving tops, wheels, and other spinning, tumbling objects have had a similar appeal for grown men and women.

Like Freud's grandson, we might not be able to master fortune, but at least we can take charge of a game that stands in for it. Marked *astragali* (knuckle bones of sheep and goats used as dice) have been found in ancient dig sites in the Mediterranean and Near East. Coin flipping was known to ancient Romans, where it was called "head or ship" (instead of "heads or tails") because most coins had an image of a boat on the other side of the emperor's likeness. An airborne, rapidly spinning coin was considered the fairest possible arbiter.

Even today, at the outset of many contemporary sporting events, teams still spin a coin to see who leads off play. In an election to the 94th District of the Virginia state legislature's House of Delegates, David Yancey, a Republican, and Shelly Simonds, a Democrat, tied with

11,608 votes each. The election was decided by a name draw from a bowl, but the candidates agreed beforehand that a coin flip would also be an acceptable method for choosing a winner.

A second, more significant use of spinning games is to open for players what have been called "portals of possibility." Such games bring about windfalls that might not otherwise come one's way. Typically, these are monetary—think of roulette wheels or the spinning drums of state lottery tickets—but not always. Take, for example, the teen-age party-game known as "spin the bottle." Most people think the game has been around forever, but written accounts of this game known go no further back than about a century. An article published in 1916 in the scholastic journal *The Playground* describes a game called "Bottle Fortune" in which players sit in a circle with an empty milk bottle in the center. One person asks a question beginning with "who?" such as "who is the most handsome person in the room?" and spins the bottle. When the bottle stops, the person it points to is declared the answer, and that person then assumes the role of spinner, and the game goes on (*The Playground* 1916).

Games like this were not just idle pastimes for bored adolescents; once, they were an important part of courtship rituals. In a culture before young men and women paired together on dates or hung out on social media, they got to know one another more casually, says Philip Conn, walking home from church, socializing at neighborhood candy-pullings and corn huskings, even, sometimes, keeping each other company through late night vigils by the side of a dead neighbor (Conn 2017).

At other times, if there was no corn to be husked or corpse to baby-sit, young people might keep each other company by playing games like "post office." Boys and girls gathered in a parlor or community building. One girl sat out of sight of the group in a different room and called out to a boy she fancied that she was holding a letter for him. To retrieve the "letter," the boy had to enter the room and kiss her.

The game of post office left no doubt as to who had a crush on whom. The game of spin the bottle, on the other hand, paired kissers and kissees entirely at random. "Spin the bottle" allows players to abandon their fate to fortune. Perhaps it's not surprising, then, that the game typically involves committing acts one might otherwise hesitate to own: *Hey, don't blame me, the bottle made me do it!* The most widely known version of "Spin the Bottle" is the adolescent kissing game in which players seat themselves in a circle and take turns spinning the bottle in the center. Whoever spins the bottle is obliged to kiss the person the bottle points to when it comes to a stop.

Middle school boys and girls have been playing "spin the bottle" at parties for almost a century (though its popularity among Gen Z has declined). But there are variants of the game for older players. By the time they get to college, the same young men and women typically substitute pints of beer for those once-novel, thrilling kisses of adolescence. Amazon advertises an electronic version complete with a blue plastic bottle that lights up when spun on its stand; "when it comes to a rest," according to the website sales pitch, "the light beaming out of the bottle cap will leave no doubt as to whom it is pointing to" (University Games 2020).

Among contemporary adult games involving spinning is the television show called "Wheel of Fortune." First airing in 1975, "Wheel" is the longest continuous-running syndicated game show in history, with more than 7,000 taped episodes to date. (Only three contestants in forty-five years have ever won the show's top prize of a million dollars, however, leading one to wonder whether American television audiences would rather see strangers *lose* money than to win it.)

"Wheel of Fortune" is an obvious instantiation of the wheel of fortune (*rota fortunae*) from medieval and ancient history as is the contemporary gaming, or roulette, wheel. Most historians credit Blaise Pascal with inventing the roulette wheel—*roulette* in French means "little wheel"—as a byproduct of his futile attempts to create a machine capable of moving perpetually. But the link between Pascal and gaming wheels is not clear, and competing theories trace the roulette wheel back to seventeenth-century English games called Roly-Poly or E-O (for Even-Odd) which were also played on horizontal, turning wheels.

I'm walking through the gaming rooms at the Circling Raven Casino Resort just south of the north Idaho city of Coeur d'Alene. The casino is located on land belonging to the Coeur d'Alene Tribe. The tribe (*Schitsu'umsh,* or "those who were found here") has been operating the resort complex ever since the spring of 1993 when Dave Matheson and the Tribal Council obtained from the state of Idaho a permit to build a high-stakes bingo hall on the grass and wetlands just outside the small (1990 population: 182) town of Worley. Ten conventional bingo games offered players the hope of walking away with as much as a thousand dollars. Most didn't—"I have to go home and tell my little dog I lost his treat money again," said Marion Preuninger, one of those early players—but no matter (Abrams 1993).

A multi-state lottery operated via telephone followed shortly, then—a testimony to people's willingness to throw away a little money on the slim chance they'll make a lot more of it—came the building of a proper, Las Vegas style casino, along with two hotels, eight restaurants, and a 7,000-yard golf course.

Around me that morning are more than a thousand gaming machines and electronic roulette wheels. It's well before noon, but the place is already packed. None of the gaming rooms have windows; the darkness, smoke, and glowing lights make it easy to lose track of time, place, even of oneself. The ceiling lights are neon, unnatural shades of blue, green, amber. Altogether a lurid scene, like something out of Ernst Kirchner's Expressionist paintings of street life in turn-of-the-century Berlin. Strolling the different rooms, I experience passing fits of claustrophobia. I consider the possibility I will never make it back to the world of grass and sunlight.

The first use of spinning wheels to play games of chance can be traced at least as far back as ancient Greece and Rome, and it's probably futile to seek a single, historic source for the gaming wheel. Some historians propose a connection between the modern roulette wheel and the round shield (*aspis*) once used by Greek hoplites. The soldiers of Alexander must have had to invent ways to fill empty hours hanging about camp, and one imagines them as no different from military personnel of any era—young men passing the time with gossip, napping, playing games or gambling with whatever equipment was at hand.

Thus, the *aspis*, a heavy (15 pounds), deeply dished shield, when it was not being used on the battlefield, could be laid flat on the ground and spun round so that it came to rest in random positions. Says gaming historian Ben Mezrich, who wrote a book about six MIT students who broke the bank at the casinos in Las Vegas: "They [the Greek soldiers] used to draw numbers on a shield, turn it on its side, and give it a spin. They would bet on which number would come up" (Mezrich 2014).

Roulette wheels and spinning bottles are basically latter-day *aspides*; both are "random number generators" designed to produce a sequence of outcomes—numbers or positions—having no discernible pattern. Modern, electronic gaming wheels come close to perfection in this respect—that is, if your ideal form of "perfection" is utter randomness—but this wasn't always the case. In the late nineteenth century, an engineer from Yorkshire turned his mechanical expertise to the roulette wheels of Monte Carlo. Joseph Jagger was his name, and textiles were his trade. (Yes, he was a distant relative of his more famous namesake, Mick.)

Perhaps because of his engineering background, or perhaps because of his observation that the wheels manufacturers used to spin cloth never turned perfectly true, Jagger assumed that the mechanics of roulette wheels in the gambling casinos of Monaco would be no different. Inevitably, he supposed, over the course of hundreds and hundreds of separate turns, imperceptible but inherent imbalances would cause the casino's wheels to come to rest at some few numbers more often than all the others. Acting on this belief, Jagger first spent a few weeks crowdfunding, dunning friends and relatives for cash. In one letter preserved by his great-great-great-niece, the historian Anne Fletcher, the gambler begs a friend, "Can you lend me five quid?" (Utton 2018).

Once he'd amassed a suitable grubstake, Jagger (accompanied by his oldest son Alfred and a nephew, Oates) traveled to Monte Carlo where the men spent the first several days tallying which numbers on which wheels came up more frequently. The variations they discovered were slight, to be sure, but they seemed to be consistent. Confident that he could rely on a particular bias he had noted in one wheel in particular, Jagger started laying down bets and winning a healthy number of them. After a few days the casino caught onto Jagger's method and began swapping wheels and tables secretly and randomly, but not before Jagger and associates had collected more than two million francs. Anne Fletcher celebrated his exploits in *From the Mill to Monte Carlo: The Working-Class Englishman Who Beat the Monaco Casino and changed Gambling Forever.* You can look it up.

CHAPTER 2

Around the House

Salad Spinners, Wet Dog Shakes, and the Rolling Pin of Doom

Long before armed civilians stormed the Bastille on July 14, 1789, an earlier, much less violent revolution had taken place in the kitchens of King Louis XIV. In the summer of 1651, François Pierre de la Varenne, the chef to the Sun King, published a cookbook. Like the *Declaration of the Rights of Man*, *Le Cuisinier français* too altered the face of the French landscape. It was, writes historian T. Sarah Peterson, "the cookbook that changed the world" (Peterson 2006). With breezy authority, Varenne overturned more than half a millennium of inherited kitchen practices, recipes, ingredients. Into the dustbin went saffron, cumin, nutmeg, and ginger; in their stead on the kitchen counter appeared common, regional herbs—parsley, rosemary, thyme, sage. Ragoûts, liaisons, and bouillon were in; breadcrumbs, the traditional thickening agent for sauces, were tossed out the window, replaced with a simple roux made of flour and butter.

Changing the royal cuisine was only one of the culinary wars Varenne had to fight. Like Joan of Arc and Nicholas Tesla, visionaries born before their time, Varenne lacked the necessary material culture that could enable the gustatory splendors he imagined. The kitchens at the palace of Versailles were so large and located so distant from the dining hall that Varenne's mouth-watering dishes often arrived cold by the time they reached the tables of King Louis. Even more challenging, Varenne wrought his culinary magic using only the most basic, hand-held implements—an array of knives, spoons, forks, a few wooden pins—tools of the cook's trade that hadn't changed much in a thousand years. To whip up his aery bisques, meringues, and Béchamel sauces, the master chef would typically brandish nothing fancier than two forks held together to make a primitive whisk.

Many of the household tools and furnishings that we use daily haven't changed their designs much since antiquity. Pharaohs of the 18th dynasty drifted into the afterlife on upholstered couches looking not much different from a La-Z-Boy. The *klismos* Plato sat on had four legs, a woven seat, a curving back; the chair looks as familiar, as chic today as it must have in fourth-century Athens. We still dine at tables supported by legs, sleep with our heads cushioned by pillows stuffed with down. But when I stand in my own middle-class kitchen, I can lay my hands on an array of cooking implements that would surely have made François Pierre de la Varenne's eyes dazzle with envy and pleasure. Within reach, resting on open shelves or tucked out of sight in drawers are two hand-held electric mixers, a food processor, two blenders (one immersion, one countertop), a dough-maker, a coffee grinder, an ice-cream freezer, and a salad spinner. Every one of these miraculous gadgets works its alchemy by spinning. Let's peek inside the cabinet drawers and pantry shelves in my kitchen, and I'll tell you about a couple of them as we go along.

We'll start the tour with one of the simplest of spinning kitchen conveniences, the hand-cranked salad spinner. The salad spinner works courtesy of Newton's second law of motion. It's constructed from two lightweight plastic bowls, one inside the other, both covered by a lid. The inner bowl is full of holes, while the walls of the outer bowl are solid. A crank handle on top of the lid lets you spin the inner bowl round while the outside bowl remains stationary.

Here's what happens when you put soggy handfuls of lettuce and arugula into the inner bowl and then spin it very rapidly. Once the water-covered leaves are set in motion, they naturally want to keep moving in the same direction, sliding across the bottom of the bowl. But the leafy greens can't go very far until they run smack into the vertical sides of the bowl.[1] Once the leaves bump against the walls of the bowl, of course, they will have no further place to go. But they're still trying to move, pressing outward; the faster the bowl spins, the harder they will be squeezed. If the walls of the bowl were solid, the result would be a limp, soggy clump. But because the sides of the bowl are full of tiny holes (like a colander), the water has an option that the leaves don't. Unlike the vegetables, the water is liquid. It can break into droplets small enough to pass through the holes and outside the bowl into the outer, solid-walled bowl where they can be collected and drained away. What's left behind after a half-minute or so of rapid spinning is an inner bowl full of mostly dry greens that can easily be fluffed out.

Getting rid of unwanted water on lettuce by spinning it away seems typical of the smarts that are characteristic of humans. But furry animals have long been using the same rotational kinetics, literally spinning water off themselves for millions of years. Four-legged creatures can't use towels, of course, and so early in their evolutionary development they had to figure out a way to dry themselves off when they got wet. This wasn't just a matter of personal comfort; in cold climates, keeping dry has always been a life-or-death matter. At some point in deep time, therefore, all mammals learned to dry their fur by executing a series of rapid half-spins, the messy, noisy, back-and-forth maneuver known as "the wet dog shake."

The animals are performing the same trick as the salad spinner, but the physics of their behavior is surprising enough to catch the eyes even of research scientists. David Hu, a professor of biology and mechanical engineering at Georgia Tech in Atlanta, has gone so far as to build a "wet dog simulator" (really!) to test the rotational forces dogs and other animals produce whenever they shake themselves dry. His findings are extraordinary. The smaller the creature, says Hu, the smaller the radius of its spin maneuver, therefore the smaller the rotational speed it develops at its surface.

Because the tangential velocity of something moving in a circle is directly related to its distance from the center of rotation (according to the formula $V_t=\omega r$), the hairs on an animal whose body is less than half a foot in diameter (like a cat, for example) will develop tangential velocities only a fraction of the speeds of bigger animals like wolves and bears. Hence very small animals must spin back and forth extremely fast if they are to generate rotational forces sufficient to rid themselves of excess water. As size goes down among furry creatures, spin velocity goes up. Large animals like bears, Hu observed, "shake at four times per second, dogs at four to seven times per second, rats at 18 times per second, and mice at a dizzying 29 times per second" (Hu 2018, 2).

The cook's spinner and the wet dog shake are simple applications of a spinning mechanism known technically as a centrifuge. Spinning a mix of different things to separate them can be just as useful a procedure in multi-billion-dollar nuclear enrichment plants as it is in household kitchens: just as you use a Mouli mixer to spin water off lettuce leaves, you can use a Zippe gas centrifuge to separate the rare and valuable uranium-235 from the heavier and much more abundant uranium-238. (The numbers 235 and 238 give the sum of the neutrons and protons in the nuclei of the two respective isotopes.)

When a centrifuge starts to spin rapidly at high speeds, it temporarily creates an "effective gravity" on whatever happens to be inside it. This force can be much more powerful than the everyday gravity we move in, as great as 50,000 times the real thing in some extreme cases! The denser bits of whatever materials the centrifuge happens to contain, whether droplets of water or atoms of U-238, are pushed more quickly to the outer edges of the cylinder than less dense materials such as lettuce leaves or atoms of the fissile U-235. Then, once water, plasma, or U-238 have been separated out, they can be collected and discarded, leaving behind substances suitable either to serve on a dining table or to install deep inside the guts of a nuclear reactor.

When they're neither drying off lettuce for a salad niçoise nor collecting enriched uranium for a nuclear warhead, centrifuges perform all sorts of innovative and highly useful tasks. In chemistry labs in high school and college, my lab partners and I took small glass tubes containing solutions of different materials and hung them from the arms of a small carousel. As soon as the machine began to spin round, the rotational acceleration swung the tubes out almost horizontally. After a minute or two of spinning rapidly, layers of yellow or green sludge at the bottom of the tubes revealed whether lead, iron, nickel, or some other element had been hiding unseen in the murk. A smaller version of the same spinning device is used widely in medical laboratories for multiple tasks—to separate whole blood into its different components, for example, or to purify viruses or isolate nucleic acids.

More recently a tiny, hand-held centrifuge has been used on an experimental basis to shortcut the time required for diagnosis of urinary tract infections. In a recent paper published in *Nature Biomedical Engineering*, researchers describe how a tool not much bigger than a fidget spinner can be used to capture any bacteria present in urine. A small amount of urine is placed in a chamber in the center of the spinner; the walls of the chamber are covered with a filtration medium. Spinning the device flings the urine against the filter which traps any bacteria that happen to be present in the sample; then, adding dye to the mix indicates the approximate level of bacterial load. The test takes less than an hour from start to finish; the researchers claim it is highly accurate and can be used to eliminate unnecessary prescriptions of antibiotics along with the long-term risks to society at large that come with overuse of such medicines (O'Grady 2020).

If you want a good feel for what it's like to be a drop of water zipping around a salad spinner or a cluster of *Escherichia coli* whipped around in a laboratory centrifuge, next time you're at a theme park or carnival, buy ticket on a ride machine called the Tilt-a-Whirl. These vomit-making whirligigs have been making riders dizzy and green since 1927, when they were first manufactured for the Minnesota State Fair by Herbert Sellner of Faribault, where they're still made today. The Tilt-a-Whirl, says Sellner's great-granddaughter Erin Ward, has always been especially popular among young kids, who for some reason seem to crave life on the edge of nausea (Haga 2001).

Much less enthusiastic, however, are military pilots and astronauts-in-training who are obliged to put in time on the 20-G Centrifuge at the Ames Research Center in Mountain View, California. After he'd had a few spins on NASA's version of Sellner's Tilt-a-Whirl, John Glenn reported that the machine was "sadistic." Same with Michael Collins, the pilot who circled the spacecraft Columbia around the moon while Neil Armstrong and Buzz Aldrin hopped about on the lunar surface. Collins called the centrifuge "diabolical." If the tortured expression on Roger Moore's face in the James Bond film *Moonraker* is any indication of what it feels like to ride such a machine, the astronauts' testimony seems accurate.

A salad spinner is a simple implement, but an even simpler spinning chef's tool is a rolling pin. Any culture that uses dough develops ways to flatten it for baking or drying. Doubtless the oldest and simplest way to get dough suitably thin for baking flat goods like cookies or pie shells is to squish it with your hands, but unless you've got the touch of a skillful baker this leaves the dough lumpy and too irregular to bake evenly. A better way to flatten a lump of dough is to spin it. This is the method still used by pizza makers.

You can watch it happening in restaurants where customers have a view of the kitchen. First the chef takes a handful of dough, squishes it slightly to form a disk, then tosses the dough into the air. Just before release, a wrist flick sets it spinning rapidly, like tossing a Frisbee. Anything that spins feels powerful inertial forces that try to stretch it outward. Those rotational forces don't significantly change the overall shape of hard, solid objects like baseballs, spin tops, or Frisbees. But in the case of a substance like pizza dough which is both sticky and elastic, they do. As soon as the spinning clump of dough is flung into the air, it starts to stretch outward, extending further and further away from the axis of rotation. As long as it's spinning, the dough will keep stretching and thinning simultaneously until—if the chef is skilled and sufficiently flamboyant—gravity pulls it back in hand as a pan-sized circle of pizza dough.

For kitchen klutzes, a rolling pin accomplishes the same stretching and thinning with rotational kinematics of a different sort. Instead of causing internal forces to stretch the dough, the rolling pin uses static friction with the dough to flatten it. Rolling motion involves rotation about an axis that is not fixed. When a car moves down the road, its wheels spin round and move forward simultaneously; they have both angular and linear (translational) velocity. Unlike sliding friction that occurs between surfaces that rub against one another, rolling friction is caused by the deformation of the two surfaces in contact with one another.[2] The point of contact between the tires and the road is always changing, and the weight of the car is always pressing straight down. As the car moves forward, its weight is distributed among the tires that push down on the surface of the road, while the road (and the ground underneath it) push right back.

Anything that happens to get between the two surfaces, of course, will feel both the force of the car pushing down and the force of the road pushing up; as a result, it will deform. Something very hard like a small rock will be deformed hardly at all, but anything soft will tend to squish and bulge out. In his book *Flattened Fauna*, biologist Roger Knutson documents in deadpan style what common species of birds and mammals look like after they've been repeatedly flattened by hundreds of rolling wheels. A rolling pin performs a similar operation on a lump of dough, flattening and stretching it like a squirrel mashed beneath an eighteen-wheeler (Knutson 1987).

Rolling pins date back thousands of years. Kitchens seem never to have been without them. Carved on the surfaces of pillars inside the necropolis at Caere, some thirty miles north of Rome, are bas-reliefs depicting daily life in an Etruscan home (c. fourth century BCE). Among the carvings in stucco are shields, weapons, chairs and couches, animals, and familiar kitchen implements for making dough: a jug, knife, ladle, a rolling pin. In many kitchens in Italy today, writes Anna Del Conte, "you would find almost identical equipment" (Del Conte 2016, 22). Rolling pins, like spinning wheels and butter churns, became associated historically with women and women's work. Their trail leads not only to kitchens and cooking but to stock comedies of domestic life and the "rolling pins of doom" wielded by angry women chasing husbands for their latest transgressions. Wilma of the television cartoon series *The Flintstones* occasionally threw a rolling pin at her husband, Fred; the humorist Ogden Nash wrote a poem about a cartoonist whose life was a never-ending stream of comic-strip fiascoes until he died when his wife did him in with a rolling pin; even in the present day, in the perennially popular Ukrainian opera *Zaporozhets za Dunayem* (*Cossack Beyond the Danube*), the scene in which a wife chases her husband with a rolling pin still draws hearty laughs (Lisovskaya 2015).

The rolling pin in my own kitchen drawer spins a maple wood cylinder around a central pin; the short, knurled handles on either side don't turn along with the roller. They're just there to hold onto as the pin rolls back and forth. This two-piece construction wasn't invented until 1884, when "J.W. Reed" applied for (and received) a patent on her "dough kneader and roller." J.W. Reed is believed to be Judy Reed,[3] who some say was the first African American woman to receive a patent for an invention; Reed's device used two rollers that simultaneously mixed and flattened dough as it passed between them. Until Reed thought to spin the wooden cylinder on a stationary axis, rolling pins were made of one solid piece of wood, tapered at each end to fit more comfortably in the palms of bakers' hands.

First generation rolling pins were little more than rods, about the thickness of a drum major's baton. Some were stout like clubs, with the ends were tapered to give a better grip; this made them resemble the wooden belaying pins round which sailors used to tie ropes to anchor the ship's sails. (The resemblance between the two implements is why they're called rolling "pins.") After a little practice, cooks learned to cradle the tool in the palms of their hands with exactly the right downward pressure to keep it rolling over the dough smoothly and evenly, flattening and stretching it while simultaneously not scraping through it or bunching it into clumps.

Like so many ancient tasks involving the hands, the act of rolling dough must have been simultaneously mindless and artful, like knitting or brushing. Most early rolling pins were shaped on a lathe; hardwoods such as cherry, maple, ash, and lignum vitae made tools that resisted moisture and cracking. When germ theories of disease took hold during the nineteenth century, glass and porcelain pins proliferated in the kitchens of Europe because their surfaces were apparently impervious.

Modern rolling pins sold in high-end kitchen shops return the tool to its historic roots. It's not just a case of retro-chic. Many cooks say one-piece designs give them a better feel for the dough. This makes sense. With the one-piece style, says Micheline Maynard, "you feel much closer to the action. Your hands become part of the rolling pin, and smoothing out the dough becomes a more sensory experience" (Maynard 2020). A one-piece pin transmits the force of rolling contact directly from wood to hand. For a cook shaping a lump of dough, the sensation would be analogous to a mechanic's feel for torque on the head of a bolt. Steel and dough talk back to you; something in the nervous system makes it feel as if the tool belonged to the hand itself along with fingers and thumb. "The hand is the visible part of the brain," Immanuel Kant once wrote, echoing the pre–Socratic philosopher Anaxagoris, who had perceived that the brain is in the hand and the hand in the brain.

Made of glass, ceramic, or composite materials, in addition to wood, these kitchen tools are both functional and beautiful as if they had been created by artists of the Weimar Bauhaus. In the centuries before mass-produced objects, rolling pins made in Shaker communities also mixed utility with art (as did so many of the tools and furniture the deeply religious Shakers used daily). Bored New England sailors carved fancy pins from exotic woods and whalebone to give to their wives on their return. At home, manufacturers began producing hollow glass pins to sell to sailors who were too lazy to carve their own. Adorned with promises like "Be true" or "May the eye of the Lord watch over you," these non-functioning keepsakes could be hung in ships' cabins or filled with salt and given as gifts. The pins could be closed with corks and hung near a fire to keep the salt dry. Glassworks in Bristol, Stourbridge, and other British manufacturing centers made rolling pins unique to the region. Bristol was famous for its deep, cobalt-blue color; glass made in Nailsea, on the other hand, was dark green splashed with white or striped in pink and blue. Some rolling pins even crossed the border from functionality into utter uselessness: painted, even gilded, such glass pins were called "love tokens" and sold for display, not intended for baking at all.

American households typically contain a dozen or more large and small electric appliances; they don't cost much in comparison to most necessities of modern life, but their effect on the way we live has been profound (Bowden and Offer 1994, 725). Common kitchen tools like mixers, food processors, immersion blenders, bread machines, and range hood fans carry out their "final cause," as Aristotle might have put it, by applying the spin of electric motors directly to blades that mix, chop, blend, suck, or blow.

Still other appliances use electric motors indirectly. In a refrigerator or freezer, for example, an electric motor spins a compressor that liquefies Tetrafluoroethane gas, then pumps it through a series of coiled tubes where the liquid absorbs heat until its molecules are again sufficiently energetic to return to a gaseous state, at which point the cycle repeats. In a small device like a coffee maker, a small electric motor spins a pump to create a vacuum that "lifts" boiling water to the top of the machine so the scalding water can then drip down through a basket of coffee grounds.

We'll take a closer look at what goes on inside electric motors a little later. For the moment, here's a quick summary of how they work. One part of the motor consists of a permanent magnet of some sort; often it's glued or screwed to the case. The other part is made of bundles of wire wrapped tightly around a spindle or shaft; this second part typically fits inside the permanent magnet where it's free to turn round. Whenever I flip the switch on my blender or clothes dryer, electricity begins to stream through the coils of wire wrapped round the spindle. The flow of electrons creates a magnetic field around the spindle which engages with the permanent field outside it. *That* field (called the stator) can't move. But the magnetic force in the stator *can* push (or pull) against the inner field of the rotor, causing a torque that makes it pivot. Inner and outer magnetic fields alternately pull and push against one another, spinning the shaft of the motor and whatever blades, pulleys, or gear wheels are attached to it.

A Rub-a-Dub Here, a Rub-a-Dub There

Midway between the salad spinner and NASA's Mother of all Tilt-a-Whirls is one of the most common and most useful centrifuges of all, the modern automatic washing machine.[4] One easily forgets what a luxury it is to put on clean clothes each morning; until the twentieth century, most people conducted business wearing the same garments that had covered their bodies the day—if not days—before. Smelly socks and stale undies used to be the normal state of affairs, morning, noon, and night. Gerald of Wales wept for his countrymen who went to bed wearing "the same clothes which they have worn all day, thin cloak and a tunic"; in Britain and continental Europe through the Renaissance, your washing machine was likely the nearest river (Harvey 2020).

Pliny the Elder (first century CE) gives specific information on the laborious procedures involved in cleaning soiled clothes at the height of Roman civilization. His *Natural History* (77 CE) is marred occasionally by tales of cyclopes and one-legged Indians who used their lone, large feet as sunscreens, but his description of washday in Rome is straightforward. The process, called fulling, required laundry workers (commonly slaves) to perform multiple steps during which dirty garments were soaked in water, pounded, and then stretched out to dry. Cleaning and restoring the nap on woolen garments also required chemical treatment with ammonium salts, one handy source of which was human urine.[5] Inevitably this practice led to jokes in the works of playwrights and other satiric writers. Here's the comic dramatist Plautus depicting a typical wash day in ancient Rome, where laundry workers, called *fullones*, commonly pissed in the wash water to help bleach out stains: "Unless I send the pimp a present today, tomorrow I'll be drinking the stuff fullers use."

Seventeen centuries later, things weren't much better in America. In 1828, the British traveler Margaret Hall wrote of American children whose clothes were "dirty and slovenly." The clothes of adults, it seems, were just as noxious, even at the highest social levels. Harriet Martineau's travels took her to one of President Andrew Jackson's raucous "levees" where she mingled with "men begrimed with all the sweat and filth accumulated in their day's—perhaps their week's labour." That most Americans of the time were dirty and smelly, however, wasn't really their fault. They just couldn't help but smell bad, says Jack Larkin, author of *The Reshaping of Everyday Life 1790–1840*, a history of daily life in America during the first several decades of the then new nation. It would have been "virtually impossible," Larkin says, for any Americans of that era to have kept their bodies and clothes decently clean by modern standards (Larkin 1988, 162–63).

When it comes to cleaning clothes, not much changed from Pliny's day until 1858, when Hamilton Smith of Pittsburgh assembled the first rotary washing machine. In his application for a patent, he touted the invention as follows: "my improvement consists in a slatted or perforated reel ... arranged to revolve within a body of water contained within an outer vessel or reservoir, and operating in conjunction with an internal weighted reel or roller ... so that the pounding of the inner reel or roller on the clothes may have the effect of forcing the water ... through the interstices of the fabric and through the perforations of the spaces of the revolving reel into the outer vessel or reservoir ... so that the latter may serve as a receptacle for the heavy particles of dirt, which are thus prevented from returning to the folds" (Smith, patents.google. com).

Smith's washing machine spun round with a hand crank, but otherwise it used the same basic physics and technology as a Whirlpool manufactured yesterday. Peek inside the machine in your laundry: you'll see either a spindle or shallow disk sitting at the bottom of a large drum. The cone or disk will have several vertical fins molded into its surfaces; the drum, in turn, has dozens and dozens of small holes covering its sides. The fins and holes in the drum are there for the same reason as the roller and perforations Smith describes in his patent application. The fins spin back and forth, hammering the dirt out of the wet clothes; when they're clean, an electric motor spins the entire drum rapidly for several minutes, squeezing the clothes against the holes in its sides through which the dirty water can be flushed away.

Credit for thinking to attach an electric motor to a rotary washing machine is sometimes given to Alva Fisher, an engineer working for the Hurley Electric Laundry Equipment company of Chicago, who took out US Patent 966677A in 1910 for a "drive mechanism for washing machines" (Fisher, patents.google.com). But in his patent application, Fisher claims not to have invented the machine itself, merely a way to drive it, and there is controversy over who should receive accolades for first reimagining wash day.

It's a cluttered field of candidates. In 1907 Louis P. Willsea filed an application to patent his "centrifugal machine" but this device, the inventor claimed, was "intended particularly to extract the oil from oily waste" (Willsea, patents,google.com). (In his application, Willsea did note that in a pinch, his "centrifugal extractor" could double up to wash dirty clothes.) There are still other prior claimants (Maxwell 2020). A hand-cranked drum machine for cleaning clothes was patented by the American James King in 1851, and still earlier in Britain, a patent was issued to Henry Sidgier in 1782 for a washing machine that also worked by means of a rotating drum.

In retrospect, pairing a rotating drum wash machine with a spinning electric motor seems a no-brainer. But the marriage was a long time coming, and on the way from Sidgier to Fisher there were some intriguing side roads contemplated but not really taken. One of them belonged to William Clack of Prairie du Chien, who filed a patent in 1871 for a wooden tub mounted atop a pair of wheels and axle. Clack's little trailer could be filled with water and soap and attached to a wagon or Sunday brougham, then tugged along on the way to church, the market, or grandmother's house, churning dirty clothes along the way. Even cleverer than Clack was Sarah Sewell who in 1885 built a rotary washing machine (US patent 3300626A) powered by see-sawing children, whose limitless energy and endless fascination with bobbing, rolling, and jumping up and down, Sewell must have reasoned, might as well be put to practical use. The device, in the inventor's words, "provides amusement and recreation for children and young persons, while at the same time [utilizing] their exertions ... in washing the family or other clothes" (Pappalardo 2019).

Early generations of electric washing machines cleaned clothes but left them sopping wet. Some of this water could be eliminated by passing the clothes through a mechanical wringer on top of the machine. From pre-school days I remember seeing in my grandparents' basement a barrel-shaped washing machine with twin rolling pins perched above; wet, clean clothes were cranked between the rollers, and the rest of the drying took place outdoors in fair weather. On rainy days or through the winter months, the clothes were hung inside on collapsible wooden racks.

When Maytag added "spin cycle" to its lexicon sometime in the late 1940s, the development of the modern washing machine was complete. Spun at about 1000 rpm, soggy garments are forced outward toward the walls of the perforated drum. Pressed tight against the sides, like the lettuce leaves in a salad spinner, the clothes are much too big to move outward any further. Not so the water in them, which is gradually pushed out of the clothes and through the scores of little holes in the wall where it's collected and drained away. Spinning at its highest rate, the machine can remove more than half the water from a tub full of soggy shirts, socks, and undies.

A Carpenter's Brace, a Circular Saw, a Record Player

The earliest tools were clubs and stones—simple, found implements used for pounding, cutting, slicing. Yet primitive, nomadic people from time to time must have needed also to make holes in things. The first crude tool for hole-making was doubtless some kind of awl created from a readymade object with a hard, pointed tip—a sharp, wooden splinter, a jagged piece of flint, perhaps a large canine tooth or tip of an antler. Soft materials like animal skins could simply be poked through. Tougher substances like wood could be punctured by twisting a sharpened tip back and forth long enough to make a hole. Among the objects perforated by hunters and gatherers of the Upper Paleolithic (25,000–12,000 BCE), however, were shells and teeth pierced for stringing to be worn as decorations. Hammers, stone flakes, or antlers could not make holes in substances like these. To make small holes in hard materials, one needs a spinning, cutting implement of some kind—a drill.

Each generation lays down technological sediment, but near the bottom layer of hole-making tools is a remarkable implement called a bow drill. This is a highly sophisticated tool that in the hands of a skilled operator can spin a slim, vertical shaft at speeds approaching 1,000 rpm. It works like this: first, a bow is strung loose enough to provide slack to loop once round a sharpened spindle and drill bit. The bow is held horizontally, the spindle and bit upright over the spot to be drilled. The worker holds the spindle and bit upright in one hand while at the same time applying moderate downward pressure; the other hand rocks the bow side to side so the looped cord grips the spindle and bit, spinning them rapidly back and forth.

Fitted with sharp, pointed tips made of chert or lapis lazuli, even the most primitive such bow drills were capable of boring holes through wood and even through much harder substances like shells or stone. The tool was used widely throughout dynastic Egypt; multiple bas-reliefs and paintings depict carpenters and bead-makers using bow drills equipped with arrow-shaped bits, and Victorian archaeologists unearthed large building stones from the Old Kingdom (c. 2500–2100 BCE) that contained perfectly circular holes apparently made using copper tube drills. The drills used to make these holes ranged in diameter

Image from the tomb of Rekhmire, High Priest of Annu (Heliopolis), 1504–1425 BCE. The scene depicts a worker (figure on the right) using a bow drill to make holes in beads for stringing (figure on left) (Nina M. Davis, Metropolitan Museum of Art).

from a quarter inch to almost half a foot, and their bits were fitted with jeweled points to make holes in limestone, sandstone, even in granite. Elegantly and precisely made, these primitive drills could have spun their way through just about any substance on the planet (Gorelick and Gwinnett 1983).

The bow drill is a sophisticated piece of technology, a huge improvement over simpler drills that were spun by rubbing a shaft and sharpened bit back and forth between the palms of the hands. Even more amazing, then, is that the bow drill is among the very oldest of all human tools. These drills appear almost universally distributed among widely different, prehistoric cultures; because of their ubiquity, says the late Professor Steven Vogel of Duke University, the small handful of cultures that lacked them attract more curiosity than those that did. Bow drills have been unearthed at Paleolithic dig sites dating well before the agricultural revolution; ten thousand years later, sophisticated versions of the same basic tool could be found in Victorian woodworkers' shops and tool catalogues. It was not until the tail end of the nineteenth century that bow drills were finally nudged aside by portable drills that spun using the then new, widely available electrical power.

Corded drills, cordless drills, drill presses, hammer drills, impact drivers, air angle drills, core drills, drills to bore ever-so-carefully through skulls or to penetrate the titanium-hard enamel of teeth. (Before the use of air-driven, high-speed drills became widespread during the middle of the twentieth century, my dentist told me, it was common during routine cavity repairs to see smoke coming off the patients' teeth.) Seed drills, water well drills and oil drilling rigs, and a tunnel-boring machine named Bertha that played a tune for workers while she slowly (at 1.5 rpm) chewed a 58-foot hole for the four lanes of State Route 99 that run beneath downtown Seattle (Szondy 2017). What philosopher, what mystic, five thousand years ago, could have guessed there were so many means and needs just to drill holes in things?

The craft of Hephaestus the metallurgist has been practiced for more than ten thousand years, but the tool commonly used for drilling holes known as the brace and bit did not come into use until the Middle Ages were seriously on the wane. A brace is a crank-shaped device that turns a spindle to which is attached a sharp, hole-cutting metal tip. The torque, or twisting force, that you can apply to turn something like a screwdriver or drill is directly proportional to the distance from the center of rotation,[6] so turning a sharpened drill bit with a horizontal crank provides much more force to drive the bit through the wood. A painting done by the Flemish artist Robert Campin in the early fifteenth century shows Saint Joseph the Worker about to make holes with such a tool in a tablet-sized piece of wood (Albury 2015). To use it, Joseph would have gripped the knob with one hand and twisted the brace round and midway round to make the sharpened metal drill bit sink ever deeper into the wood.

Another early example of a brace and bit comes from the wreck of the British warship *Mary Rose*, sunk in 1545. Found in the ship's carpenter's chest, the brace drill used on the flagship of King Henry VIII consists of a wooden, C-shaped crank with a knob that spins freely at one end and a hole to attach a cutting bit at the other. One more historic drill comes from the fourteenth-century castle Merwede in the Netherlands. This tool was once freeze-dried to preserve it; unfortunately, the method didn't work—freezing made the wood much worse—and the second oldest carpenter's brace in history has now become unusable and is sadly disintegrating.[7] At least the part that has crumbled away opens an inside view of the simple mechanism for attaching a knob handle to the crank. The knob fits loosely on the crank handle, and a wooden pin extends through it and into a groove cut into the handle to secure knob to crank while allowing the crank to spin a drill bit continuously.

The Merwede brace is a simple-looking, even crude, tool. But if you were to set it next to the brace and bit on my workshop table the kinship would be obvious: each tool consists of a large, C-shaped crank handle with a loose, freely rotating knob on one end and a drill bit attached firmly to the other. The only real difference between these two tools half a millennium apart is in the shape of the bits themselves. The contemporary brace is fitted with an auger, or screw bit. This type of drill bit was a great improvement over bits shaped like spoons or arrows, shapes such as were used by Saint Joseph or the carpenter on the *Mary Rose*. The deeper that spoon- or arrow-bits dig into a piece of wood, the more difficult it becomes to clear the growing pile of chips and waste shavings out of the hole.

An auger bit, on the other hand, drills and cleans holes simultaneously; it's modeled after the water delivery system of ancient civilizations which used large, rotating screws to raise water up from rivers or dug wells. Instead of lifting water from a river, the auger drill bit lifts wood chips from a hole. This both keeps waste material from clogging the drill and makes drilling cleaner and faster. As the screw turns, the newly sliced chips climb up the rotating slope of the auger just as smoothly as one might ascend a steeply spiraling staircase.

When a brace and bit lie on a carpenter's workbench, or hang from a peg on the wall, they give the space around them a kind of authenticity. From the tool emanates the aura of a trade and the labor of hands that have used it. On the bench before me is just such a set. Before the days of portable, 20-volt hand drills capable of boring all day long through floor joists and framing studs, if you wanted to make a hole in a piece of wood, you needed a hand brace much like this one. The drill bits I bought myself years ago, but the brace once belonged to my grandfather, Edmund Hartman; I carry his name between my first and last. It's got to be more than a century old. A wooden head and handle are worn smooth; the grain is open and dark. The metal on the spindle and chuck is pitted, but the manufacturer's stamp of identification is still legible: "Stanley Tool and Level Co. New Britain, Conn. USA."

In the summer of 1972, I used Edmund Hartman's brace daily when I installed electric wiring throughout the log cabin my wife and I had moved into earlier that year. With the tool I drilled dozens and dozens of holes for Romex cable that I threaded through the 2 × 4 studs we had nailed to the logs inside the house. The studs didn't hold anything up; we used them just to make space for batt and roll insulation and to give the cabin's interior, rounded walls a flat surface. In hiding all the electric wiring inside faux stud walls, we were trying to make the house safer and neater. It would have been easier to run the cables through metal conduit, or, like the people who had lived in the cabin before us, just run wires along the logs, fully exposed. Easier still would have been to turn back the clock a hundred years and live without electricity entirely. But then I would not have had the chance to commune daily with my grandfather.

Orhan Pamuk writes in *The Museum of Innocence*: "The power of things inheres in the memories they gather up inside them." This seems not quite the whole of it. Better is this poem by Bertolt Brecht called "Fishing Tackle" ("Das Fischgerät"):

> In my room, on the whitewashed wall
> Hangs a short bamboo stick bound with cord
> With an iron hook designed
> To snag fishing-nets from the water. The stick
> Came from a second-hand store downtown. My son
> Gave it to me for my birthday. It is worn.
> In salt water the hook's rust has eaten through the binding.
> These traces of use and of work
> Lend great dignity to the stick. I
> Like to think that this fishing-tackle
> Was left behind by those Japanese fishermen
> Whom they have now driven from the West Coast into camps
> As suspect aliens; that it came into my hands
> To keep me in mind of so many
> Unsolved but not insoluble
> Questions of humanity.

You grip a tool in your hand, begin to use it, work with it. You picture the person who owned it before you, doing the same thing. It's hard not to feel the kinship, to believe that the other person is there, that you are in some way the same. Whenever I spin the handle on my grandfather's brace, I'm not just summoning up memories of lost times and people. I'm also feeling something somatic and deeply empathic—what Brecht calls the "traces of use and work." Responding primarily to the sensation of touch, the worn knob on the handle or the stubborn resistance as the bit chews into wood, mirror neurons fire off in my brain. They trigger emotional resonances; I jump the gap between me and my grandfather. I become one with him by using things he used.

The enormity of the debt we owe our forebears never ceases to humble me. When Heinrich Gruber disembarked from the ship *Dragon* at the port of Philadelphia in 1732, he gave his occupation as *Dreher*, or "turner." The German word described someone who was skilled at shaving—turning—wood on a lathe. It connoted someone highly skilled; today we would him a cabinetmaker. The lathe Heinrich used was likely a simple device, little more than a table beneath which was installed a foot pedal and crank to spin a drive wheel on top. But with his lathe he could have turned out anything from diminutive egg cups and knobs for dresser drawers on up to salad bowls, balusters, even (had the sport been invented in 1732) baseball bats.

Wood lathes of earlier centuries were still simpler machines, a couple of spindles extending inward from mounts set vertically atop a narrow platform. The "dreher" mounted a length of rough wood between the spindles so it could rotate freely. Meanwhile a helper looped a strap or cord around the workpiece, using the same basic technology as a prehistoric bow drill, and pulled the strap back and forth, first spinning the wood in one direction, then another. As the wood spun, the turner shaved off bits of wood until, as Michelangelo once said of the sculptor's art, all superfluous material had been chiseled away. It was a herky-jerky business, and the two workers had to take care to coordinate their separate tasks, like performers in a *pas de deux*. But because the axis of rotation remained constant no matter which direction the wood was spinning, in a short time a deft turner and assistant could shave a lopsided chunk of wood into a flawlessly symmetrical bowl or table leg.

I learned to use a lathe as a freshman in high school in J. Robert Gibson's shop class. I'd enrolled in basic wood shop classes the previous year, but middle-school boys were restricted to using only simple hand tools. We worked for a month with a small block plane just to square one end of a pine board, learned the difference between crosscut and rip saws, drew precise, isometric sketches of boxes we never built. The following year we practiced using power tools—table saw, band saw, planer, and the lathe. When we got the chance to turn wood on a motor-driven lathe, most of us were content to carve out things shaped vaguely like batons or cue sticks. But one boy, Robert Gerhart, had a bolder vision; he spun a chunk of walnut into the shape of a large, stiff penis. The rest of us were delighted and envious; why hadn't we thought of that? Even Gibson, a characteristically dour and terse man, couldn't stifle his amusement. There was mirth in his eyes when he spoke his obligatory, teacher's reprimand: "Oh, for Christ's sake, Gerhart!"

Editors at *Popular Mechanics* complained not long ago that "the saw that built America" was now made almost exclusively in China (Berendson 2019). Mine was, anyway. Whether they're made domestically or abroad, however, a portable, electric circular saw bids fair to be the most important hand tool of the last hundred years. A modern portable circular saw contains mostly composite- and plastic parts, but a tool manufactured yesterday still looks much like the one first created by Edmond Michel in 1923. Nobody really knows how many portable circular saws have been made since then. But if you're reading this, you likely have watched one in use or used one yourself. It's guaranteed that circular saws have been an important part of your life—especially if the house you live in or the building you work in was built after World War II. The construction boom that created the modern American lifestyle mix of suburbs, malls, and highways could never have happened so quickly and completely were it not for the portable electric circular saw.

The ingenuity of the circular saw is easy to overlook because teeth are everywhere in the natural world. They can be found on widely different things like sharks' mouths and pampas grass. Toothed blades were likely used by the earliest human cultures, perhaps because such cutting implements were so easy to adapt from nature's own creations. Even proto humans may have begun to cut things apart with the sharp, jagged edges of seashells, sharks' teeth, possibly even sawgrass. The poet Ovid tells of Talus, Daedalus' son, who as a child, "studying the spine of a fish, took it as a model, and cut continuous teeth out of sharp metal, inventing the use of the saw." Cultures able to fabricate toothed blades out of metal were much better able to thrive and to expand than were groups that were dependent entirely on whatever saw-like materials nature happened to provide.

Otherwise, there's no real difference between the edges on the teeth of a great white shark and the framing saws used to cut wood for the theater in ancient Athens. Both implements work in exactly the same way: a row of hard, small serrations—the word derives possibly from Proto-Indo-European *sers-, "to cut off"—is dragged forcefully across a hunk of flesh or a piece of oak, tearing it apart as it goes. By the third millennium CE, Romans had constructed a sawmill at the Greek city of Hierapolis in (now) southwestern Turkey to cut apart blocks of building stone. Learning to use toothed implements for cutting might not even be limited to humans. *Spy in the Wild*, an animal documentary produced by the BBC, shows an untrained orangutan using a handsaw after having observed nearby villagers building huts (Hale 2020).

In the ancient world as well as in medieval Europe and colonial America, sawing trees into square, dimensioned building lumber was historically a two-man operation. Logs were laid across a deep pit, then sawed lengthwise into boards by two men using a thin-bladed "whipsaw." One man stood on a platform next to the log and above it, while the other was stationed in the pit below. The work was slow, hard, and unpleasant, especially for the person down in the pit who was bathed constantly in falling sawdust. Even worse was the inefficiency. Anyone who has used even a common handsaw knows the intermittent, back-and-forth nature of the operation. The teeth cut only on the forward, or push stroke; the back, or pull stroke accomplishes nothing except to position the row of teeth for a subsequent cutting stroke. One full hour of sawing, in other words, is half spent getting ready to saw.[8]

It is not known who first imagined a straight row of teeth furled around the circumference of a spinning, steel disk. On the internet I found competing claims for the invention of the circular saw from seventeenth- and eighteenth-century sawmill owners in Britain, Holland,

and Germany, also from a seamstress harmoniously named Tabitha Babbitt who lived in a Shaker community in Massachusetts. In 1810, Babbitt is reputed to have mounted a circular saw blade to the hub of her spinning wheel. Supposedly, Babbitt had been inspired to create a saw that would cut continuously as she was watching a couple of local sawyers waste half their energy and time on the backstroke. Or so the story goes, anyway: Babbitt's claim to the invention of a saw that spins has been largely discredited, says the American historian M. Stephen Miller (Miller 2010, 184).

Whoever he (or maybe she) was, the inventor of the circular saw was a marvelously clever person. Where would you get the idea in the first place? It's not a design drawn from nature, where teeth always come in neat, straight rows. The circular saw basically takes that linear, front-to-back arrangement of cutting teeth and loops it round on itself to form a circle, turning a straight row of teeth with a beginning and end into a line of cutters infinitely long. Once you see such a saw, its advantage over any straight-toothed tool is immediately obvious. Because the cutting teeth on a circular saw are placed evenly and continuously on the outer edge of a spinning disc, some of them are *always* facing forward, *always* chewing through wood.[9] Forget the stuttering, off-again, on-again rasping of pit saws, crosscut- and whip saws.

Within a few decades after its invention, the high, unbroken howl of water- or engine-powered circular "buzz" saws was being heard in towns both large and small everywhere throughout Europe and America. ("The buzz saw snarled and rattled in the yard," wrote Robert Frost in memory of a New Hampshire youth who had been killed in an accident involving such a saw, "And made dust and dropped stove-length sticks of wood....")

The back story is more complete for a type of circular saw small enough for one person to operate and to carry about from one job to another. Good things come to those who wait, as the saying has it, but fortune sometimes blesses those who are hard at work looking for something else. Such was the case with one Edmond Michel and his creation of the portable, hand-held circular saw. Born in France in 1857, Michel had emigrated to the United States early in the twentieth century and settled in New Orleans; a census lists him as married and living with his wife and five children in the city's Sixth Ward. In 1921 (it is said), while he was watching local sugar farmers with machetes hack their way slowly through fields of thick cane stalks, Michel was inspired to create a powered cutting tool that was small enough for workers to carry on the job. He first bolted a two-inch, circular blade something like a pizza cutter to the front of an old machete; next, a mechanical "worm" gear meshed with a toothed wheel attached to the hub of the blade. The arrangement converted the spiraling movement of the worm into spinning motion of the blade. To power his new cutting machine, Michel cannibalized an electric motor and drive shaft from a malted milk mixer.

This tool proved unsatisfactory, mainly because the worm gear spun the round, slicing blade much too slowly to be practical. A worker using even a dull machete could cut through the cane stalks more quickly. Then, something clicked in Michel's imagination: he trashed the pizza cutter and connected the motor, worm, and drive gear directly to the back of a bigger, *toothed* blade. When he'd finished, Michel held in his hands the world's first portable, circular saw. The thing was not only faster in the sugar cane fields, but it was also sufficiently powerful, Michel soon discovered, to chew through inch-thick wood as happily as through cane stalks.

Michel immediately patented his worm-drive saw. Then, together with an investor named Joseph Sullivan, in 1923 he founded what was to become first company to manufacture hand-held circular saws. The two set up the Michel Electric Hand Saw company in Chicago, hawking the first few saws to home builders in Los Angeles and to contractors beginning work on what was then the brand new "boardwalk" in Atlantic City, New Jersey. (Portable circular saws not only built the famous Boardwalk itself, but they also built almost every structure that ever stood on it, including a casino once owned by Donald Trump as well as a temporary wooden platform set up in front of that same building where once, during her presidential campaign, Hillary Clinton railed against her opponent's "business" of running businesses into the ground.)

Builders and workers were enthusiastic about the Michel Electric Hand Saw; the future looked bright. But Michel seems to have preferred inventing to selling. He soon sold out his share in the business to his partner, not even guessing that the company he had founded would subsequently become a generic trademark for the tool he had invented: Skilsaw. Edmond Michel parted ways with Joseph Sullivan in 1926 and returned to New Orleans, his place in the pantheon of important inventors now largely forgotten. Meanwhile Sullivan's fortunes prospered: Skil Corporation (now a subsidiary of Chevron HK Limited) is the world leader in do-it-yourself tool sales, and old Joseph's great-granddaughter, Jenny Sullivan Sanford ("Mother. Author. Leader," as she describes herself on her personal website) made a career in investment banking.

One spinning thing you won't see in my house—or most any contemporary American house for that matter—is a record player. The phonograph used to be ubiquitous; like the radio, with which it was sometimes paired in sleek wooden cabinets, these sources of music and information once rivaled kitchen tables as centers of family gatherings. To have a console unit on display in the parlor was a kind of badge of social identification; especially in America, where social position was not always easy to signal by way of dress or accent, to own a Magnavox or Zenith stereo console signified a certain level of both income and taste.

As young parents during World War I and through the 1920s my grandparents had run a small-town bakery, but the Depression cost them first their business and then nearly their home. To get to the other side of the 1930s and then through the war years, my grandfather took part-time work for the roads department of Whitehall Township in southeastern Pennsylvania, spreading crushed rock and tar on potholes and setting out rows of flickering smudge pots to warn nighttime motorists of road hazards. In their house I remember seeing only a rounded, table-top radio and a record player housed in a small, fabric-covered suitcase. The phonograph was especially fascinating because much of its complex arrangement of levers, spindles, and rotating disks was exposed for a five-year-old to contemplate. The turntable revolved at a single speed, 78 rpm, which was fast enough to fling off dice and dominoes that I dropped on the outer edges of the black, spinning disk. A decade later in my parents' house we heard music and the occasional news broadcast on a radio and turntable that were hidden inside a gleaming mahogany console the size of a large chest freezer.

Record players. Tape decks. Compact discs. Spin the dial on the face of a radio; as a child I listened, rapt, while snippets of music and static buzz drifted in and out of focus as the needle swept over the band width. Once, Americans' households contained so many of these devices that conjured music and voices from the empty air, it seemed impossible to wrap your mind around their wonderful variety. Now they are antiquated, has-beens—jokes, really. Measured in years, it didn't take long for their time to pass. And what was a record player, anyway? Some clunky, retro machine that once, in the middle of a nationally televised debate, an embarrassingly enfeebled presidential candidate summoned up from some fond, childhood reverie as a thing that might bring back harmony and mutual respect to family life.

I'm old enough to forgive Joe Biden for believing that families ought to gather daily around a record player. The technology of the twentieth-century record player, as every school child once knew, was the brainchild of Thomas Alva Edison. It blends physics, harmonics, and electromagnetism. The first phonograph captured sounds on a cylinder wrapped with metal foil. A needle attached to the narrow end of a bell-shaped tube pressed against the cylinder which could be turned round by hand. As Edison spoke into the tube, sound waves from his voice were transmitted down to the needle which absorbed them, vibrated in response, and cut sympathetic grooves in the foil of the revolving cylinder. The result was a physical imprint—a record—of the sound of Edison's voice; all he had to do to hear repeated verbatim what he had just said was to reverse the process.

When he ran the needle along grooves that had already been cut, it vibrated in the same pattern as before. Now, however, instead of recording sound vibrations, the needle obligingly sent them back into the tube where they were amplified and played back as something that could be heard. Edison spun his original phonograph (from Greek *phonē*, "sound," and *graphē*, writing) with a hand crank; similar machines powered by small electric motors came onto the market after World War I. They captured sound for playback on flat disks instead of cylinders. Still, like Edison's original machine, these "record players" used a stylus that bumped obediently through every nook and cranny in grooves cut in a circle on the surface of a hard disk. (Early disks were made of a shellac-based material and were quite fragile, easily broken; all-but-unbreakable vinyl disks came later.)

The last record player I owned was portable. Buttoned up inside a livid clamshell case, it traveled with me first to college in the sixties, then found a place ten years later in the bedroom of my pre-school daughters who listened for hours to the ethereal voices of Joan Baez and Judy Collins or to Disney cartoon movie soundtracks—*Snow White*, *Sleeping Beauty*, *Cinderella*. My own mix of records—Chopin's nocturnes and Beethoven's symphonies rubbed dust jackets with Peter Paul and Mary, Edith Piaf, and Tom Lehrer—went unheard for decades. I lugged scores of those and other similar albums in three fraying cardboard boxes from house to house until one summer afternoon in 2012 in suburban Atlanta when I set them by the curbside for trash pickup the next morning. Mysteriously, the boxes were gone by nightfall. It would be wrong to say that I miss the sound of a phonograph record, the background hiss and static crackle; I'm content now to summon Alexa to broadcast whatever song or singer I fancy. Still, in that part of my heart where my children are forever young, I hear Elaine and Laura scuttling about their bedroom, now and then pausing to call out to me: "turn record over."

CHAPTER 3

Amazing Grace

Spinning Poets, Whirling Dervishes

The story of how poems are made has not changed in two thousand years. From Plato's *Ion*: "the poet is a light and winged and holy thing, and there is no invention in him until he has been inspired and is out of his senses, and the mind is no longer in him [but] simply inspired to utter that which the Muse impels." I tell the myth to undergraduates at Emory University in a class called "Introduction to Poetry." They are respectful but disbelieving; how quaint, how silly this seems. Many of them will soon enroll in courses where they will be instructed how to write stories, plays, poems. Creative writing is a thriving industry on American campuses; each year, MA and MFA degree programs churn out as many as 4,000 graduates. Is it possible to teach someone how to write a poem? Or is poetry, when it comes, unruly and unbidden, a gift from the muses on Mount Helikon? Here is singer Alanis Morissette, speaking in an interview for MTV's Week in Rock on August 20, 1995: "The songs in this set were all written in about 15 minutes. Many times, it felt as if it was all being channeled through us."

Rumi, the thirteenth-century Sufi Muslim poet and mystic, is said to have composed his verses in a state of ecstasy brought on by the repetitive beating of drums or the sounds of a favorite water mill. One day, strolling the bazaar in Konya (near modern Ankara), Rumi happened to hear a goldsmith beating metal inside his shop. Like a tuning fork, he responded to the blows of the hammer, whirling to its rhythms in the street. Later, he will tell followers that when choosing a road that led to God, he took the one paved with dance and music.

Assembled from stories and pieces of stories, a portrait of Jalāl al-Dīn Muhammad Rūmī, the spinning poet, shows a boy who on the same day played street games with his peers after roaming (or so he claimed) the stars and planets in the company of angels in green robes. Rumi (a nickname meaning "the Roman") is said to be the best-selling

50

poet in the United States.[1] His philosophy has inspired New Age writers like Deepak Chopra; the British actor Tilda Swinton borrowed some of his verses to promote her line of perfumes; Madonna credited his poems with inspiring her album "A Gift of Love" (1998); and his verses, many of which are said to be suitable for framing, do a brisk online business:

> Dance, when you're broken open.
> Dance, if you've torn the bandage off.
> Dance in the middle of the fighting.
> Dance in your blood.
> Dance, when you're perfectly free.

While living in Konya—newly married, preaching, further pursuing his scholarly career—Rumi one day met a strange, rambling mystic named Shamsoddin, "Shams" for short. The gospel Shams broadcast was stern and sere; to him, words were veils, barriers to insight and understanding. In place of language and conversation Shams recommended music, song, and above all a particular kind of whirling, spinning dance. This last form of worship, Shams said, he had learned from dervishes lodging with the Sheikh of Tabriz. The encounter with Shams was life-changing for Rumi, to say the least. Immediately after their meeting, he abandoned his duties at the madrasa where he had been teaching and moved into Shams' rented room at an inn used by merchants, where for the next three months (in the words of Brad Gooch, one of his biographers) he was taught to "spin loose of language and logic, while opening and warming his heart" (Gooch 2017, 124). What emotions, what lore and learning passed between the two men during that time, one can only guess. But their relationship must have been as complete as it was profound: no sooner had Rumi returned to his family and students than he took a second "sabbatical," this time living with Shams for six months.

So close, so fierce a bond could not last. There was friction, there were arguments, words that could not be recalled. At length the two broke. Shams departed abruptly, without notice, and when he came to understand that his friend was likely gone from his life forever, Rumi, according to his son, "roared like thunder" (Gooch 173). In his grief he began to compose poetry and to practice the mystical, whirling dance that he and Shams had once shared, rejoining his beloved friend by way of empathic, somatic modeling. For many years after his death, townsfolk of Konya still spoke of a former teacher who one day had become demented, speaking "only in Persian rhymed couplets which no one could understand" (Gooch 173). Rumi's son, Sultan Valad, in his *Book of Walad*, recalled when Rumi "danced his repetitive spinning to music long into the night.... Day and night he began to dance *sama*, on the ground like a spinning wheel" (Gooch 175). The poet's sanity, his spiritual revival, his verses, all seem to come from the eerie, whirling dance.

For hundreds of years a group of dancers known as the whirling dervishes of Turkey have been commemorating this 800-year-old Persian poet by donning long skirts and, like him, spinning themselves like tops. The spinning of Sufi dervishes bears only a superficial resemblance to the artistic spins of skaters and ballerinas. Unlike ballerinas or hip-hop artists, the whirling dervishes have no extensive repertoire of dance moves to speak of; they just spin round and round while their billowing skirts ripple and flow in ever-changing patterns. Yet the dancers achieve beauty almost beyond imagining. Aesthetics may not be their aim, but their performances bloom with sensuousness. Like ballet dancers for Edgar Degas, the spinning Sufi worshippers have inspired painters like the Pakistani artist Shafique Farooqi, who says of his more than 10,000 works: "My art is deeply inspired by the whirling dervishes at Rumi's mausoleum. I have tried to express his thought by painting whirling dervishes" (Shafqat 2020).

Literature on the whirling dance is sparse but rich. Perhaps this is because the dance is so full of contradictions: stasis vs. movement, ecstasy vs. insight, holding tight and letting go, losing in order to find. There was great curiosity about Persian mysticism in eighteenth-century Britain, and the strange, whirling dancers sometimes caught the eyes of painters and poets. The British artist William Hogarth, for example, is best known for his moral works, in particular the sequence of eight paintings known collectively as "A Rake's Progress." Hogarth's canvases depict the stages in the life of a young spendthrift who squanders his fortune on parties, gaming, and sex, ending finally with his imprisonment and confinement in Bedlam, London's infamous mental asylum. On the flip side of one of these "pictur'd Morals" (a phrase the actor David Garrick composed for Hogarth's tombstone) is an engraving from very early in the artist's career, a sketch titled "The Inside of a Mosque, the Dervishes Dancing" (1724).

Hogarth, of course, had seen neither the inside of a mosque nor whirling dancers in person; his view of the whirling dancers draws on paintings by Jean Baptiste Vanmour and Gerard Jean-Baptiste Scotin as well as narrative accounts from a contemporary travelogue, Aubry de La Mottraye's "Travels throughout Europe, Asia, and Into Part of Africa." In the engraving, Hogarth depicts five Sufi dancers in traditional long skirts and peaked hats. He collapses different elements of the dance into a single moment; three dancers spin with their arms extended and heads cocked sideways, one kneels, and a fifth crouches low to kiss the tiled floor. The scene is three removes from reality, a picture of a picture

"The Inside of a Mosque, the Dervishes Dancing," an 18th century engraving of spinning Sufi dancers by the British artist William Hogarth (1697–1764). Hogarth depicts a passage from the French traveler and diplomat Aubry de La Mottraye's travelogue in which the author describes a ceremonial performance of the Mevlevi Sufis, popularly called "whirling dervishes" because of their constant rotational movement (William Hogarth, Metropolitan Museum of Art).

of a picture: Hogarth modeled his work after an engraving by Vanmour ("The Dervishes Turning in their Mosque at Pera"), which Vanmour had based on his earlier *Collection of One Hundred Prints Representing Nations of the Levant* (1714).

Searching online for some images of this legendary dance performance, I open some YouTube videos of contemporary Sufi performances. Even shrunk in size to fit onto the screen of an iPhone, the whirling dancers are mesmerizing. Heads canted, eyes open and cast upward in an empty, thousand-yard-stare, arms fully outstretched yet not stiff, the dancers spin slowly counterclockwise. With each revolution, they balance on one foot and push off with the other; the motion is somewhat like that of a skateboarder cruising slowly down a sidewalk. I would not dare to pretend I can put myself in the shoes of these dancers. Still, I can sense their potent spirituality: they blend color, movement, the physics of divinity and grace. The dance also mixes sensuality with abnegation, self-indulgence with asceticism.

I come across a web-only magazine called *The Gilded Serpent*; for more than twenty years, the publication has been advertising itself as the "journal of record for Middle Eastern Music, Dance, and Belly Dance." Contemporary dervishes (from Turkish *derviş*, a religious mendicant) who carry on the tradition begun by Rumi, writes Cara Tabachnick, fast for hours before the ceremony. A performance begins, she says, when the dancers begin to turn "in rhythmic patterns, using the left foot to propel their bodies around the right foot with their eyes open, but unfocused.... As the dancers turn, the skirts of their robes rise, billowing into circular cones, as if standing in the air on their own volition" (Tabachnick 2019). The skirts' lower edges are spun outward at rates up to four times the accelerative force of gravity, creating ripples and sequences of waves, according to Tabachnick, "which seem to defy gravity and common sense."

Dozens of images and informational videos of actual or imitative Sufi dancers can be viewed online. The dervishes characteristically whirl at about 60 rpm, one full turn for every second. A dancer rotates simultaneously around two different axes, turning around himself while also wheeling slowly around the entire circle in conjunction with the rest of the performers. From the point of view of basic physics, the dervishes carry out two different versions of rotational motion, each dancer rotating individually at the same time the entire group translates round and round like the buckets on a Ferris wheel.

Because Islam forbids dancing as worship, reports Roberta Strauss Feuerlicht in *The New York Times*, the performers of the Mevlevi order in Turkey avoid controversy with a fine parsing of terms: they claim not to be dancing as such but merely "turning" (Feuerlicht 1975). Whether dancing or turning, performers complete only a single, repetitive step, a basic shuffling, shoving pivot while standing flat on one foot. Gracefulness does not seem to be a prerequisite for success on the dance floor. (This indifference might have appealed to me as a clumsy adolescent when I was made to run a gauntlet of complicated dance steps from waltz to fox-trot to tango. What would not have appealed, however, was the method with which the whirling dance is learned: beginners practice by placing their big and second toes around a large nail driven into the floor, chanting as they turn.)

Recently the artistry of the spinning dervishes has attracted curiosity not just among tourists but among physicists. One factor of interest seems to be the haunting, rhythmic motions of the dancers' skirts as they rise and fall, ripple and flow. It seems that the dancers really are in tune with the universe. A few years ago, a group of international researchers looked at the mechanics of Sufi whirling and discovered that it mimics some of the largest spinning forces on the planet, in particular the Coriolis Force. Equations published in 2013 by the Institute of Physics and German Physical Society show that the patterns that appear and dissolve in the folds of the dancers' robes follow closely the development and disappearance of large flows of air in the Earth's atmosphere.

The same planet-sized rotational forces that spawn tornadoes and tropical cyclones, in other words, account for the sensuous, rippling flows of individual Sufi skirts. Just as the spinning of the Earth deflects things moving across the surface counterclockwise in the Northern Hemisphere, clockwise in the Southern, the constant spinning of the dancers causes rotating, pyramidal designs to form in their long, flowing skirts. "The flow of a sheet of material is much more restrictive than the flow of the atmosphere," says James Hanna, one of the co-authors of the study, "but nonetheless it results in Coriolis forces. What we found was that this flow and the associated Coriolis forces, plays a crucial role in forming the dervish-like patterns" (*Science Daily* 2013).

The dancers intend their spinning to function as a kind of meditation; unlike the playful twirling and rolling kids perform on the playground, Sufi whirling is not supposed to make you dizzy. On the contrary: the sole purpose for the dancer, in the words of one participant in a recent dance workshop held at the Menla Retreat Center in upstate New York, is to use the spinning human body to "reverse engineer the entire cosmos" (Crittenden 2017).

Sufi whirling has some emphatically mundane offshoots. Give people something to do, anything at all, and chances are they'll make a competition out of it. Residents of Tokyo flock to see the annual Naki Sumo Baby Crying Festival, and back in 1999 a bored garment worker named Philip Leicester convinced some friends to join him in a sport he called "extreme ironing" in which contestants took an iron and board to a dangerous and (preferably) remote place to press the wrinkles out of some clothes (Banu 2018).

It's no different with Sufi whirling, where records have been set and record-breaking attempts continue to be made. *Guinness World Records* states that the current record for consecutive rotations (both sexes) in one hour is held by Nicole McLaren who spun without toppling for 3,552 turns. On the way to establishing her record, McLaren spun at an average 60 rpm—one complete rotation circle every second for nearly an entire hour. As for the category of "most people Sufi-whirling simultaneously," Guinness awards that honor to the 755 dancers who spun at National Taiwan University on November 20, 2011, in support of the Taiwan Cancer Friends New Life Association (guinessworldrecords. com 2016).

Dance Moves

Horses on medieval battlefields were performing pirouettes hundreds of years before there were ballet dancers, keeping their hind legs more-or-less in one spot while using the front legs in a series of side-steps to turn a complete circle. For horse and rider to master this swift, spinning maneuver was essential in combat, as the pirouette was the only way for a lone warrior to thwart an enemy attack from behind. For a dancer, fortunately, the stakes are not so high, and the pirouette is executed not in self-defense but in the spirit of exuberance, for the sake of beauty. Repetitive turns have long been among the repertoire of basic moves for any dancer. Performing spins in ballet probably began as a variation, or spontaneous improvisation, upon the broad circling and twirling movements that are as old as human dances themselves. Requiring a dancer to rotate the body over a single supporting foot on the floor, one or multiple times, the pirouette is perhaps the most familiar spin in classical ballet.

The success of a balletic pirouette depends entirely on the performer making use of the invariant mechanics of spinning bodies. To perform a pirouette requires a whole-body rotation while balanced only on a pointed toe; the dancer moves her center of mass from support on two legs to balancing on only one, while simultaneously executing one or more rotations. Since postural sway is a fundamental component of any human movement, the most difficult aspect for dancers who attempt to complete multiple pirouettes is to maintain a true vertical body alignment through each successive spin. This looks hard, and it's even harder to do than it looks.

One research article I consulted allowed dancers a deviation from true verticality of no more than a single degree over successive spins! For a dancer who is five and a half feet tall *en pointe*, that's a margin for error of about an inch. Slim margin indeed.[2] Outside that narrow range, she will wobble like a toddler. To maintain that vertical spin axis takes extraordinary discipline and strength; as a professional dancer and choreographer, Sally Radell, put it (in something of an understatement), "Ballet is really hard." (Professionals often make what they do look easy to naïve spectators; country singer Dolly Parton, famous for her gaudy outfits and exaggerated facial make-up, once told her audience, "You wouldn't believe how much it costs to make me look this cheap!")

Skillfully executed, even this most basic dance move is of surpassing grace; the turn requires the dancer to stand on one leg held vertical, while the other (or "gesture") leg is raised with its foot alongside the knee of the supporting leg. As with almost any rotational dance move, the pirouette begins when the dancer shifts her weight to one foot and pushes with the other against the floor simultaneously up and sideways, creating a torque that produces a rotational acceleration. The greater the force of the push-off, the greater the angular acceleration that causes rotational motion. Once the gesture leg has pushed off and lifted, of course, the dancer has no other source of torque, hence no way to acquire more angular momentum. But even though the laws of physics prevent her from gaining additional momentum once she begins to spin, while she is turning, she can at least maximize the effects of whatever angular momentum she does possess by maneuvering different parts of her body.

For example, it is common for a dancer to extend her arms in the direction of the turn when first pushing off into a spin. The outstretched arms then carry a quantity of her initial angular momentum which can be "conserved" in the form of an increased spin rate as they are tucked in closer to the dancer's torso. It looks magical, but it's not at all illusory. The conservation of angular momentum increases the spin rate of the dancer's body, just as a diver rotates faster as she curls into a tuck, or as the winds inside a rotating funnel cloud increase as it stretches skyward.

The pirouette is sometimes credited to a German dancer named Anna Friedrike Heinel, who starred in Parisian balletic performances in the middle of the eighteenth century. (Men perform pirouettes too, of course, but commonly they spin while standing on the ball of one foot rather than on *pointe* which of course makes it much easier for them to maintain balance and a vertical postural line.) Another possible creator of the pirouette is Marie Allard who is known to have created the pirouette *a la seconde*, a spin in which the raised leg is held at a right angle to the supporting leg. Most dancers perform the spin move in sequences; performing four or five pirouettes in succession is common, and modern dancers have executed as many as fourteen of them in a row.

I talked with Radell, chair of the dance program at Emory University in Atlanta. Radell has been teaching and performing for more than thirty years, and dance has been at the center of her life for as far as she can look back. "I remember when I was six, we lived in Winnetka, Illinois," she told me. "We had one of those big lumbering TVs, and it was a Sunday morning or something and I was sitting on the ground in my pajamas, you know, eating graham crackers and milk. I was just kind of sitting there with my blanket, just hanging out, and some ballet came on. It was a romantic ballet, a fantasy scene, and I thought oh my God! It was the most beautiful thing I had ever seen. They were doing pique turns and shimmering, and I just thought: *Okay, that's it*. And I remember making a promise to myself that this was the best, the most beautiful and exciting thing I had ever seen, and I would do whatever it took to get there. It was as if I made this commitment with God. All of a sudden, you're touched, and you know what you need to do."

From that moment a young girl took herself, her life in hand. "I remember trying to make my own pointe shoes," Radell said. "My dad had a wood shop downstairs. And I went down and I found two chunks of wood, maybe two by two, and I had a little sewing machine. I found some fabric, and after I was supposed to be in bed, I would take the fabric and wrap it around my feet with the wood stuck in the fabric over my toes. There was a radiator in my room behind the door, and I would stand behind the door on my fake pointe shoes and hold my hands on the radiator, like it was a ballet box, until my feet were in terrible pain, and I went back to bed."

Radell talked about the struggle to rid all traces of self-consciousness from the dance. "If you can do a clean pirouette, then you work on a doubles, and then you work on triples until everything kind of comes together. You know, dancing is an addiction. It's mental up to a degree, and then it's unconscious."

"You mean like what they tell a baseball player? Full head, empty bat?"

"Yes, you want to get it to that place. You keep doing it over and over and over again until you've patterned in just the right relationship. Another thing is that as a turner or a spinner usually one side is better than the other. I can turn to my left much better than to my right. I always loved it when we turned on the left side because I could get that one. Usually, dancers turn to one side one side better than the other. And I think it must have to do with handedness, like my turns on my left side were more together than my turns on the right."

Multiple variations of the pirouette exist. Other spin moves that are part of the classic ballet repertory are several variations on the pirouette such as the *grande pirouette,* in which the dancer turns with the gesture leg pointed outward at 90 degrees, or the arabesque turn. And then there's the most difficult spin move of them all, the turn known as the *fouetté.* The fouetté—a French word meaning "whipped"—requires the dancer to raise one leg, extend it, and snap it around to the side, causing her body to revolve on her other foot, *en pointe.* She whips the raised leg sideways to provide torque to set her body in motion. What that leg is doing is creating and momentarily storing momentum. Then, as she lifts herself up on point, that leg is drawn inward, and the principle of conservation of angular momentum carries her through the rest of the turn. Lift, snap, and spin. Lift, snap, and spin. Repeat as necessary.

To perform a single fouetté is difficult; to perform thirty-two of them in succession, as is required of Odile, the treacherous, black swan maiden in Pyotr Ilyich Tchaikovsky's *Swan Lake*, all the while never wandering or falling, is something of a miracle. Even very great ballerinas regularly substitute less challenging maneuvers for Odile's fouettés; sometimes they omit them entirely. So famously difficult are these spins that there's even a website dedicated to their performance: 32fouetts. com.

Writing for *The New York Times*, Alastair Macaulay says that Odile's 32 spins may not be the most challenging maneuver in ballet, but they're without doubt the one with the greatest potential for embarrassment. Anyone who performs them is totally exposed to failure. It's like an outfielder who drops a fly ball; when you muff it, there's nowhere to hide. "If something goes wrong," says Macaulay, "the audience will see. I remember one ballerina falling flat on her backside around turn No. 14 and several who stopped (or switched to another step) after about 20" (Macaulay 2016).

I asked Radell whether she thought about what her performances look like to the audience? Not at all, apparently. Onlookers were not on her mind, she said, because the dancer herself was in a state of mindlessness.

"I don't think about what I want the audience to see," she told me. "Because that takes energy away from my execution. There's a sense when you get almost to the point where you feel like you're a machine."

"O body swayed to music, O brightening glance/How can we know the dancer from the dance?" So, there it was again, one of the great riddles of physical performance: at its highest levels, humans measure themselves against some non-living ideal. Mechanical mothers proved to be catastrophically incapable of nurturing infant monkeys that were placed in their care. But ballet dancers, like professional athletes who sometimes talk of "being in the zone," often describe their most effective performances as robotic. A ballerina spins *en pointe* mindlessly and without any perceptible sway; she strives for a mechanistic standard, to become a top. Indeed, the word "pirouette" comes from the middle French *pirouet*, or "spinning top," a Latinate formulation which in turn has roots in the older Roman *pirolo*, a child's top.

Doubtless the performer who first spun round on one leg (Heinel? Allard?) reminded onlookers of the familiar, mechanical plaything. Ballet inspires awe not just because of the beauty and grace of the dancer's performance but also because the physical movements themselves seem to go beyond the limits of biology, crossing over into an eerie, non-living state. Because an element of postural sway is physically essential to every motion that involves legs, arms, and torso, a perfectly executed sequence of pirouettes showcases a living/non-living paradox. It takes naturally fluid human movement and measures it against something—a spinning top—utterly incapable of it.

On Behalf of Recess

> [A]s yet I am stupefied & don't know how to be suffi-
> ciently thankfulilified, that I can do nothing but turn
> round on my toes like a pivot.—EDWARD LEAR, in a letter
> to his childhood friend, Fanny Coombe

Children adore putting their bodies through movements that would take adults to the brink of nausea, sometimes past it. Spinning is especially popular among kids in elementary school, as are rocking, tumbling, swinging, and even swaying. According to researchers at the Penn State Extension, behaviors like these are not only normal, but they are also desirable and possibly essential if kids are later to feel at home with themselves as grown-ups. Spinning in circles, they say, "is one of the best activities to help children gain a good sense of body awareness" (Penn State Extension 2017). Vertical spinning is said to release pain-suppressing beta-endorphins and to promote right/left brain synthesis (Wallace 2004). "Through spinning, kids figure out where their 'center' is." The same rapid whirling that would topple grown men and women makes kids more sure-footed; by spinning they discover a more acute sense of bodily "centeredness" which they then integrate with most other physical movements and activities.

A hundred years ago, manufacturers of playground equipment routinely pitched their product as much for its moral as recreational or physical benefits; an advertisement from the Studebaker Park Combination in 1926, for example, claims that swing sets and merry-go-rounds were a vital part of children's future as citizens: "[b]etter the playground and trained leaders than reformatories and uniformed guards" (Rieger 2018). Nobody now expects playgrounds to substitute for family, school, or church. But there are still opportunities for kids to spin themselves senseless at recess. Some of the newer devices are a giddy whirligig built for two, an "integration carousel" with accommodations for wheelchairs, and a hemispheric, rubber-lined cradle called "the nest" that spins like a gyroscope on its pedestal (Link 2018).

Spinning fast not only benefits kids to be nimbler afoot; making like a top quite likely brings with it cognitive benefits as well. In a book called *Astronaut Training: A Sound Activated Vestibular-Visual Protocol for Moving, Looking and Listening,* researchers Mary Kawar and Ron and Sheila Frick argue that kids need opportunities during the school day to perform repetitive, mind-numbing spins and rocks and swings that drive adults crazy (Kawar, Frick 2005). The title of their book may be clunky, but they make a cogent argument to include *more* time for recess in the school day, not less.

The rotational forces kids feel when they're spinning like tops open a dialogue between brain and body; whether children know it or not, say D. Angelaki and J.D. Dickman, by stimulating the vestibular system they're also "developing the skills necessary for future tasks like following lines of text across a page" (Penn State Extension 2017). It's an argument to bring back old-fashioned merry-go-rounds and tall swings that in the interests of safety (not to mention lawsuits) have all but vanished from school playgrounds.

The children's therapist Angela Hanscom, for example, concurs; she draws a direct link between the increasing numbers of school kids with poor attention spans and the replacement of old-style swings and roundabouts with modern, safety-oriented playground equipment that doesn't spin kids woozy. Pediatric therapists, says Hanscom, currently use swings and other devices to stimulate the vestibular complex found in the inner ear. The physical activity, she says, helps children be better students, "to improve self-regulation and sustained attention to task." "As a therapist," she says, "I believe the merry-go-round is one of the most powerful therapeutic pieces of playground equipment ever invented" (Strauss 2017).

Souls on Ice

Spinning on skates makes use of the same physics of gravity, torque, and angular momentum that apply to ballet dancers. Figure skating, however, unlike ballet, grows from utilitarian origins when centuries ago it developed as a practical mode of wintertime transportation in northern latitudes. People have been navigating frozen lakes and rivers on some form of ice skates ever since the Bronze Age (third to first millennia BCE), mainly because skates offered by far the easiest way—indeed, often the only way—to travel even short distances during harsh European winters (Heidorn 2020). As anybody who's ever seen an animal try to walk across an icy pond can attest, feet—even four of them—are useless when the coefficient of friction at the surface is an impossibly slick 0.05. (A pair of Nikes on a concrete sidewalk is about twenty times stickier.) Sliding along on narrow blades made of deer and elk antlers or metallic alloys is the only way to travel in these conditions; the thin, hard, sharpened contact surface cuts into the surface of the ice and makes it possible to move forward while at the same time staying upright.

The transformation of a handy, mundane skill into a sport or an art form did not occur until medieval times. Early skating competitions were mostly speed races staged in the Netherlands in the fifteenth-century; events such as figure skating, in which skaters were judged according to how accurately they traced a sequence of compulsory geometric figures on the ice, were not part of skating competitions in Europe until the late eighteenth century. As for events where skaters perform artistic jumps, dance steps, and spins before a panel of judges, these were added to international competitions scarcely more than half a century ago.

Still, despite the lack of historical precedent, it is easy to believe that a few of those Bronze Age skaters must have been much better than their companions at gliding from place to place on slivers of horn or bone, and it is also easy to picture these more accomplished skaters occasionally performing leaps and twirls in the long, cold Nordic twilight just to show off their prowess or, like a bird that suddenly bursts into song, simply out of pure *joie de vivre*.

"When you land that first jump," says Stephanie Rosenthal, "it's indescribable. Nothing in life ever has ever felt like that." Rosenthal first put on a pair of skates at age five; she's been performing axels and salchows, sit spins and camels for more than a quarter century. She performed in the opening ceremonies at the Salt Lake City Winter Games in 2002, skated as a stunt double in the Disney movie, *Go Figure*, and won a standing ovation at the 2006 United States Figure Skating Championships at the Savis Center in St Louis. Since Rosenthal retired from competition, she's been coaching younger skaters much like herself, just a few years ago.

And then there are the spins. "[J]umps look like sport," says Christopher Clarey, but "spins look more like art. While jumps provide the suspense, spins provide the scenery" (Clarey 2014). Spinning is an important part of both types of skating maneuvers—lutzes, axels and salchows include one or more airborne rotations—but some of the spins performed by figure skaters are ends in themselves. Spectators typically see these spins as less difficult than the spectacular leaps; to uninformed eyes, they look like "breathing points" or "transitions" to more difficult parts of a skater's program. But their flawless elegance belies their rigor.

Ronnie Robertson, winner of a silver medal at the 1956 Olympics in Cortina d'Ampezzo, Italy, was said to have spun so fast that he ruptured blood vessels in his hands. Or consider the career of Lucinda Ruh. Ruh, who grew up and trained in Japan but skated in competitions for Switzerland, never found her way to the platform where medals were bestowed on the top Olympic skaters. Yet her status among figure skaters is the stuff of legends. She's still known as "the queen of spin."

On April 3, 2003, Ruh's spinning earned her a spot in the *Guinness Book of World Records*. Actually, two spots: she first gained entry for spinning 105 complete rotations while balanced on a single foot, a record which she herself broke later that day when she spun 115 times. She could have added a third citation when she spun at six rotations per second (or 360 rpm), except that Guinness monitors informed her they had "no category for this feat" (Ruh 2011, 208). But such glory came with a price. Now, years after her days of competitive skating are over, Ruh still suffers from random blackouts and chronic dizziness; physicians trace her afflictions to a lifelong history of "mini concussions" resulting from the extreme rotational forces generated when she was spinning. "I had been spinning at least two to three consecutive hours a day," she recalls, "and many more hours on some days, every day since the age of four. Just spinning and spinning in one direction"(Ruh 2011, 208).

"When I used to spin," Ruh wrote in her autobiography, "time stopped. It had no significance." Her description of the act of spinning sounds eerily like the meditative state sought by Rumi and the Sufi whirling dervishes. "Spins were to be done at times like a prayer," she said. "[I] tried to spin myself into a 'blissful, trance-like meditation'" (Ruh 2011, 177, 4). Rosenthal's take on spinning is not much different. Whenever she was spinning on the ice, Rosenthal said, she entered a trance. "I would go into it and wouldn't even feel like I was spinning because I was spinning so fast."

In kinesthetic terms, the spins executed by a skater resemble a ballet dancer's pirouettes; both performers have to hold a true vertical line to keep from wandering or precessing. Hard as that is for a dancer, it's even harder for a skater, says Rosenthal, who has experience both on ice and at the barre. A ballet dancer standing on point can bring her weight down on a single spot, but a skater is not balanced on a pointe shoe but on a thin, curving blade. Spinning on ice has one clear advantage over whirling *en pointe*, however: the medium itself. This is because the surface on which a skater spins is extremely slippery—almost free of friction, in fact—so that once anything is set spinning on ice there's not much can slow it down. An object spinning on ice will retain almost all of the angular momentum it has to start with, and that means that any changes to its overall shape will result in significant changes to the speed with which it rotates.

A skater begins to rotate by extending her arms, digging the teeth at the front of one skate into the ice, then pushing off hard with the other skate to provide an initial torque to start her body rotating. At first, she spins at a relatively slow rate, perhaps once every second or so. But as she draws in her arms and tucks up one leg, unexpectedly she begins to spin more rapidly. Her body somehow rotates faster and faster, drawing energy from an unseen source as if she were a runaway truck or a turbine spooling up.

To spectators, it may look unnatural, even eerie, but the skater's rapidly increasing spin rate is perfectly consistent with the law of the conservation of angular momentum. The product of a spinning object's moment of inertia (the distance between its spin axis and its center of mass) and its tangential velocity must remain constant. By pulling in her arms, therefore, a skater moves a significant part (about ten percent) of her overall body mass closer to her center of rotation. This simple gesture instantly reduces her moment of inertia by a factor of two while simultaneously increasing an initial spin rate of, say, one revolution every two seconds to a burred finale six times as fast—three spins per second, nearly 200 rpm.[3]

Rosenthal still talks about spins with enthusiasm; of all the maneuvers in a skater's repertoire, spins are closest to her heart. "When you go into a spin, you draw into yourself," she says. "When I'm in a spin I can't really see outside, I just let the world pass by. You feel the acceleration, but you're not thinking about it, you're thinking, 'am I centered?'" Those hours spent spinning on skates came in handy for her now and then as a college student. At undergraduate parties at Yale where alcohol flowed freely (often too freely), Rosenthal said, her friends sometimes groused about the dizziness they experienced when they'd drunk unwisely. Most people who consume alcohol to excess hate the vestibular disorientation that accompanies drunkenness. They resist the dizziness; now and then they freak out. Not Rosenthal, though: "When I would get the spins," she said, "I would lie down on my bed and just enjoy it. It felt comfortable to me. It didn't feel overwhelming. I could just watch the world go by. It was like, *Oh, yeah! I have the spins!*"

Tiny Dancers

A tennis ball dropped onto an asphalt court bounces straight back up; it won't quite make it all the way back to its starting height, of course, because some of its kinetic energy is lost to compression, friction, and sound. Otherwise, it obeys simple Newtonian physics: the downward force of the ball is matched perfectly by the upward force of the ground. It's the same with a falling drop of water. When it lands vertically on a hard, flat surface, the drop simply splashes radially. Catch one in a high-speed photograph just at impact and you'll see its kinetic energy distributed evenly in a lovely, liquid crown.

Now imagine playing a tennis match where any given serve or ground-stroke combined the slow, high-kicking bounces off the red clay at Roland Garros with the flat, lighting-quick skids on the slick lawns of Wimbledon. Such were conditions created by Dr. Yanlin Song and his research team at the Chinese Academy of Sciences when they were studying how falling drops of water react with different surfaces they happen to land on. As part of their investigation, Song's team first covered small metal plates with a fluorinated, nonstick coating. They then masked the plates with several different designs—among the shapes were moons, pinwheels, and circles with wavy lines—and exposed them to ultraviolet light. Under the aggressive, caustic light, the Teflon coating on the unmasked regions of the plates degraded quickly, making the exposed metal surface highly receptive to moisture—"sticky," as one science writer put it—but leaving the fluorinated surface beneath the designs undamaged.

The contrast between a "wettable" surface interspersed with highly water-resistant graphics made it possible for a single drop to experience widely different molecular interactions depending on which part of the drop hit which part of the surface. When, during the team's experiments, a drop landed simultaneously on surfaces with two different coefficients of friction—one that was relatively sticky, the other much less so—the energy of its rebound was applied at ever-so-slightly different times to its different halves. The half-drop that landed on a "sticky" part of the surface rebounded more slowly than the half that met a water-resistant part of the plate. These different surface interactions caused wildly unpredictable differences in the way the parts of a single drop bounced back up. Part of the kinetic energy of the falling drop kicked in instantly off the non-stick surface, the other part milliseconds later, and these tiny imbalances within the body of the raindrop created a torque about its central axis, setting it turning and twisting above the specially surfaced metal plates fashioned in the Academy's labs.

By using different sticky/non-sticky patterns, writes Nicholas St. Fleur, "the researchers [could] essentially control the dancing drop's choreography" (St. Fleur 2019). Slo-mo videos of Dr. Song's experiment present a captivating sight. The drops spin with watery "arms" extended, gracefully spiraling upward and sideways as if they were dancers executing pirouettes or pique turns. Depending on the mix of stickiness and repulsiveness they strike, the droplets respond with a variety of artsy dance maneuvers, spinning at rates up to a dizzying 7,300 rpm. Yet these laboratory performers live and die a mayfly's existence; almost as soon as the tiny dancers appear, they vanish.

Such seemingly whimsical experiments may prove to be more than laboratory curiosities. There may be ways to turn spinning, dancing water drops into practical technologies; for example, working together, hundreds of thousands of them might someday scrub off dirty windshields or de-ice airplane wings. To be sure, we're not talking horsepower or kilowatts here—the average-sized raindrop carries only a minuscule amount of kinetic energy, on the order of one-thousandth of a joule, or about as much energy as it would take to lift a correspondingly small, micron-sized part of a hamburger off your plate (Wylie 2019; McClenon 2012). Still, research is being carried out to determine whether in regions of the earth where rainfall is predictably heavy (such as Indonesia), the kinetic energy of billions and billions of raindrops falling on panels made of piezoelectric substances could supply alternative and inexhaustible electric energy (Defiyani 2018; Wylie 2019; McClenon 2012).

But let's leave these visionary possibilities for the future. For the moment, let's just enjoy the exquisite artistry Dr. Song and his research team have wrought. The twisting, dancing droplets look like miniature stage performers, eerily alive. Incomparably graceful, they might have spun straight out of postwar Germany's Bauhaus and the theatrical artistry of Oskar Schlemmer and his ethereal, abstractionist ballets.

Interchapter I

A (Very) Short History of a Metaphor

A long time ago I enrolled as an undergraduate in a course in the history of the English language. Three weeks practicing Anglo Saxon until we could blunder through the Lord's Prayer: *Faeder ure þu þe eart on heofonum.* Two weeks on Grimm's Law, the Norman Conquest, a British wordscape suddenly awash in strange, romance sounds. Then Chaucer's poetry and the Great Vowel Shift, "beet" morphing into "bite," "boat" into "boot." Near the end of the term, we were asked to complete an exercise which seemed at the time insultingly banal for a collection of upper-class, know-it-all English majors. We were to choose any three poets we liked and to select one poem and one prose passage from the works of each. There were to be no sophisticated interpretations, no "close readings" of these texts: all we had to do was to compare the vocabularies poets used in their poems with the ones they used when they wrote prose. More specifically: what were the proportions of "native" (Anglo-Saxon) words and foreign (mostly Latinate) words in the poems, in the prose?

I'm pretty sure Professor Marie Borroff knew exactly what we would find. But for the rest of us, the results were eye-popping. I picked Milton, Poe, and Eliot, a trio I thought were about as different as poets could be from one another. Yet the etymologies of their poetic vocabularies told a different story. In their poems, all three leaned heavily on simple, everyday words with old, Germanic roots. The tallies were almost identical, close to ninety percent. Vocabularies in the prose compositions, in contrast, were much different, much more varied: broader, polysyllabic, highly Latinate. The lesson was as clear as it was intriguing: whenever English-speaking poets wrote poems, they became Anglo-Saxons.

The earliest words were nouns and verbs, simple references to matter and movement. They told of things that were seen, deeds that were done, and the oldest terms in any language record the ceaseless and varied interactions between men and women and a realm of objects. The simple verb "spin," for example, can be traced back to an ancient cluster of Indo-European verbs and nouns all related to clothes or some sort of body covering. The word and its historic antecedents (e.g., *spinnan* both in Old High German and Anglo Saxon) is based mostly on notions of "turning around" or "twisting lightly" as, for example, in braiding or plaiting. Both verbal and nominative forms of "spin" (as well as related words like "spindle"), therefore, at first must have referred to the act—or the art—of creating thread from mats of disorganized fibers. After tufts of animal hair or plant fibers were spun into much longer lengths of cordage, they could then be used to make clothing or to bind two or more objects (such as an arrow shaft and sharpened stone point) tightly together. The word "spin" is still widely used in this fundamental sense, of course, both literally and figuratively. For a few hundred dollars, one can still purchase retro, wooden spinning wheels and spindles such as Sleeping Beauty once might have pricked her finger on.

As either noun or verb, the word "spin" can be (or has been): an act or spell of spinning, whirling, circling, or tumbling head over heels; the product of a toffee-making machine; a quick trip rowing or sailing, or—in Australian slang—a piece of good or bad luck and (in the years before decimal currency) a five-pound note. Victorians applied the word to an unmarried woman. (A publication from that era, *Lays of India* by "Aliph Cheem" [a pseudonym], proclaims: "O spins! Be warned ere yet too late!"). Once upon a time the word also referenced a kind of small, bodily protuberance: a sixteenth-century translation of a German text describes the uvula as "a lytell deme hangynge in ye throte lyke the spynne."

More recently, a surprising and unusually large vocabulary of highly specific meanings associated with "spin" comes from the realm of quantum physics, including terms like "spin flip," "spin orbit," "spin-spin," and "spin wave." These and many other mostly compound words are used to describe behaviors and qualities of different sub-atomic particles that carry a mysterious, intrinsic kind of angular momentum physicists somewhat imprecisely call "spin." (More on this later.)

Applications and extensions of *spinnan* proliferated during the nineteenth century. Their marvelous range of meanings is testimony not just to human ingenuity but to the subtle and various ways in which things that spin have influenced life and thought; they represent what Professor Stephen Pinker of Harvard calls "pedestrian poetry." To go (or to take) a spin referred to travel by horse-drawn wagon or (later) in a motor car; in both cases the trope linked travel with a mental picture of rapidly spinning wheels. To "spin off" the road (and possibly careen into a ditch or telephone pole), on the other hand, described the actual rotations of an out-of-control vehicle. In the armed forces during up until about World War I, the same phrase (or a variant, to get in a spin) meant either that one had failed an examination or had been brought up for reprimand on some minor offense.

The widespread proliferation of motor cars and airplanes in the twentieth century brought even more extensions of meaning. To "spin one's gears," for example, meant any fruitless labor; it called to mind an engine and drive train spinning round furiously but not actually moving the vehicle. On the other hand, to understand how someone might "go into a tailspin" (or, in a related phrase, go into a flat spin) one needed to picture an airplane plunging from the sky suddenly and catastrophically. The first use in this last application dates from very early in the aviation era; in 1915, the British weekly magazine *The Aeroplane*, signaled the then-novel meaning of the word by placing it in quotation: "Several times their aeroplane got into a 'spin.'"

Contemporary exercise machines shaped like bicycles are marketed colloquially as "spinners" or "spinner bikes," even though they are incapable of forward motion because they have no road wheels. They consist merely of twin pedals and cranks that turn a resistance belt endlessly. But refer to such machines as "spinners" with caution. Believe it or not, a company in California (Mad Dogg Athletics) has trademarked the prehistoric words "spin" and "spinner" so that in the world of contemporary exercise machines these words (like "Kleenex," "Bubble Wrap," and "Popsicle") must now by law refer specifically to Mad Dogg's line of products. Chavie Lieber of the now-dismantled web site *Racked* complained that Mad Dogg once sent him a cease-and-desist order for using "spin" in a generic way. Lieber found it both amusing and annoying that cycle manufacturers and proprietors of indoor gyms cannot call what they make or do "spinners" or "spinning," even though these same words have been spoken and written for many thousands of years. Henceforth they are restricted to terms such as "stationary bikes" or "indoor cycling."

Whether "spinning" or "cycling indoors," there are internet classes for riders of these bikes that go nowhere; there are regional competitions, there is even "Peloton," an exercise equipment and media company whose stationary bikes take riders from all over the world on imaginary races. The company manufactures cycling machines and sells online exercise programs that permit users to log onto mobile apps where they attend motivational classes or participate in virtual races and other stationary biking competitions. Like many recent tech start-ups, Peloton (the company is named somewhat ironically after the French term referring to the bunched-together group of cyclists in an actual, long-distance touring group or race) makes most of its money from the subscriptions it sells rather than any material, durable products.

We can also "spin" facts to suit our own personal interests or needs. Spin-doctors and spin-meisters weave their deceptions in spin-rooms. "From its beginnings in the idea of honest labor and toil," says Lynda Mugglestone of Oxford College, "[spin] has come to suggest the twisting of words rather than fibres—a verbal untrustworthiness intended to deceive and disguise" (Mugglestone 2011). Such negative meanings, says Mugglestone, have developed only relatively recently in the history of English. The *Oxford English Dictionary* cites, among others, *The Guardian* (22 January 1978) which warned readers to weigh information coming from the CIA "for factuality and spin."

The connection between spinning threads and twisting words and deeds likely is older even than this, possibly deriving from the different kinds of spins that baseball pitchers imparted to balls in a deliberate attempt to trip up batters. Pitchers and cricket bowlers were the original "spin doctors," and the history of negative associations with the word spin can be traced at least as far back as the middle of the nineteenth century when the then president of Harvard College, Charles William Eliot, denounced baseball pitchers who threw the newly invented, rapidly spinning curve balls as a practitioners of a "low form of cunning."

Twisted fibers, curve balls, stars spinning slowly across the night sky, the unforeseeable turns that befall any life. That the course of human events took turns for better or worse as frequently as a single spot on a spinning wheel reversed direction was a frequent *topos* in antiquity among philosophers. It is not certain when or within which culture the image of the Goddess Fortuna arose. Perhaps the association of fate with spinning and a turning wheel was an adaptation from Greek legend of the Moirai, three goddesses who were responsible for determining the destiny of everyone ever born. Clotho ("the spinner") spun a thread representing a person's life; Lachesis ("the disposer") measured its length; and Atropos ("the inevitable") cut it off.

Or perhaps the source was Tyche, the daughter of Oceanus and Tethys, who was often pictured holding a wheel or balanced atop a ball as if to display her capriciousness and the likelihood that she could—and would—move unexpectedly in any direction. The first known reference to the wheel of fortune among Roman authors comes from Cicero in a speech (*In Pisonem*, 55 BCE) denouncing debauchery among past and present consuls: "not even then did he [a reference to Calpurnius Piso] fear Fortune's wheel [*fortunae rotam*]" (Robinson 1946). Shortly afterwards, the poet Ovid (43 BC–18 CE) wrote ruefully in a letter from exile about a "goddess who admits by her unsteady wheel her own fickleness." But the closeness in time of these references suggests that both writers were drawing on a trope that was already in common use.

Wherever and whenever its source, the image of Fortune's spinning wheel obviously took hold among Roman writers, spreading swiftly and widely like a virus. A century after its first appearance, the historian Tacitus (56–120 CE) was complaining about its banality. Despite its commonness (or perhaps because of it?), the image of the goddess Fortuna atop a turning wheel acquired its greatest influence in the Middle Ages when it appealed to Boethius (c. 480–524) who warned readers against the folly of trying to halt Fortune's whirling wheel. Chaucer, too, attributes the course of events less to human decisions or their personalities than to Dame Fortune, "who always will assail/With unwarned stroke those great ones who are proud."

Half a millennium later, before he was dismissed from Fox News because of his alleged maltreatment of female employees, Bill O'Reilly habitually ended his nightly broadcast, "The O'Reilly Factor," by gesturing with his pen in a twisting motion and assuring viewers that "the spin stops here." In fact, the spin never does stop—not on Fox News nor any place else, for that matter—and on balance that's a good thing. If we can spin yarns both literally and figuratively, we can also, like Yeats, spin tales to keep a drowsy emperor awake or, in considerably more desperate circumstances, spin them like Scheherazade to ward off dying.

In the spring of 1970, I was living in San Francisco's Mission District, and I watched with a fascination that bordered on the morbid as all around me young men and women were detaching themselves from the normal pathways of American life much as skin sloughs dead tissue. At the time, I was technically unemployed. I had just quit my job writing news and features for the *Melbourne Age*, and I had begun working a meaningless series of short-term assignments for Russell Kelly Office Services, a supplier of temporary employees who were then universally known as "Kelly Girls."

Over the course of six months, performing ad hoc tasks in downtown banks and offices, back street loading bays, and the data processing department of the Veterans' Affairs building on Polk Street, I ran across an engaging if somewhat scrofulous collection of co-workers, all of whom, like me, seemed to have been flung rudely off Fortune's Wheel. We were all in the process of dusting ourselves off and trying to figure out the next move. Truth be told, most of us were wondering whether it was worth the effort to make a move at all. It was these peoples' stories I tried to tell in a series of short vignettes called "Spinoff from the Great Society." I sent the manuscript to the usual suspects, but the piece was never published, and one year and one personal watershed later, I abandoned writing fiction, enrolled in graduate studies in literature, and never looked back. I thought at the time it was a waste of a good title. I still do. The book you're now reading is a kind of return.

Chapter 4

A Day at the Ballpark

The Hell It Don't Curve

> Physicists insist that the hop on [Sandy] Koufax's fastball was an optical illusion created by the expectation that it would drop more than it did. "Physics is full of shit," said another expert on the subject, Jim Bunning, a Hall of Fame Pitcher who graduated to the United States Senate.—JANE LEAVY, *Sandy Koufax: A Lefty's Legacy*

Imagine yourself in the batter's box as the great Yankees pitcher C.C. Sabathia rocks back and goes into his windup. The ball is completely hidden from you. Even when the six-foot seven Sabathia removes it from his glove and rears back his left, throwing arm, his hand is so large it wraps completely around the ball, making it disappear. He's Goliath on a mountain top, so close he seems ready to smite you.

Abruptly, Sabathia strides forward, left arm flying. You catch sight of a small, white object bursting out of that huge left hand. It's coming straight at you, speeding at better than ninety miles an hour. There's barely time for your eyes and brain to identify *ball!* and focus on it when—*right now!*—you must make two crucial, separate decisions. You must decide, first, whether the pitch will be a strike. And second, almost simultaneously, you must decide whether to swing at it. You must make both these decisions in two-tenths of a second, perhaps a little less. That's about the minimum time necessary, researchers say, for humans to fixate visually on something and then to form a first impression. Short as that time is, it's all the time you've got. By the time two-tenths of a second have passed, the ball is halfway to the plate.

Perhaps you've heard somewhere that playing baseball is a leisurely, pastoral experience? Not hardly. The game takes place at quantum-computer speeds near the absolute limits of human physical capabilities. For you, the person standing in the batter's box, there's no time for doubt or delay. Suppose during those first two-tenths of a second, you decide to swing. It'll take *at least* another two-tenths of a second for your brain to tell your muscles to get the bat off your shoulder and start it moving through the strike zone. Two-tenths of a second to see the ball, track it, and commit to a swing; another two-tenths of a second for your bat to sweep across the plate. That's a *minimum* four-tenths of a second for your brain and body to do everything you need them to do. It's biomechanically possible to hit Sabathia's pitch—but only just. Between the instant he releases the ball and the moment it smacks into Austin Romine's mitt, slightly less than half a second will elapse. You better not blink.

As Robert Hooke in the seventeenth century revealed with his microscope the astonishing reality of the microworld, so digital clocks and high-speed cameras capable of measuring time and change by fractions of tenths of seconds do the same thing for contemporary sports. Graphics from PITCHf/x website, for example, superimpose tracking data for Sabathia's fastball onto that of his slider. (A slider is a curve ball that swerves—or "breaks"—simultaneously down *and* sideways.) The data show that Sabathia's slider and fastball follow identical trajectories for the first two-thirds of their flight. Then, at about ten or twelve feet from home plate, the slider deflects sharply downward. You'll see the ball start to drop, and you might think to lower your bat. But you won't be able to do that. Your bat is already moving at 70 miles an hour, carrying about 8,000 pounds of kinetic energy. It's too late to change the path of the swing. In fact, it's too late for your brain to tell your body to do *anything* at all. The ball will take less than a tenth of a second to cover those last few feet, and the human body isn't built to function that fast. "When the message has reached the brain," wrote the mid–nineteenth-century scientist Hermann von Helmholtz, "it takes about *one tenth of a second*, even under conditions of most concentrated attention, before volitional transmission of the message to the motor nerves enabling the muscles to execute a specific movement" (Canales 2009, 4).

In short: it's the element of disguise that makes a curve ball so difficult to hit. Even when batters expect it, the late, dipping swoop has been fooling them for more than a century and a half. The pitch works like a blind curve on a highway that suddenly tightens up on drivers when they're halfway into it—at a point, that is, when they're already committed to a speed too fast to make it through the rest of the turn safely. Does a curve ball "fall off the table" or cascade like a "mystic waterfall" (as Jimmy Wynn once described the curve ball of Dodger legend Sandy Koufax)? Perhaps not—the editors of the old *Life* magazine were technically correct in describing the overall, sloping trajectory of even the most sharply breaking curve ball as "parabolic." But it's parabolic with a difference![1]

It All Begins with Newton

One of the first people to speculate about the sharply curving paths of spinning balls in flight was the Western world's premier observer of physical motions of all sorts, Sir Isaac Newton. He wasn't looking at baseballs, of course—they wouldn't appear for another couple of centuries, and on the other side of the Atlantic Ocean. Instead, Newton (he wasn't yet "Sir Isaac") was watching tennis balls. One day in 1672 while he was watching a match at his college in Cambridge, Newton observed that balls hit with topspin—that is, balls that spun top-to-bottom (or "forward") with the direction of flight—consistently dropped to the court sooner than those balls that spun backward or not at all. Pondering what could account for the difference, Newton guessed that an imbalance of air pressures on the top and bottom surfaces of the ball might be the explanation.

His reasoning went like this: when struck with topspin, the ball's upper side would be rotating into or *against* the air during flight, while its underside would be spinning away from it; hence the top of the ball would be moving faster relative to the oncoming air than would its bottom surface. The top of the ball would meet air at a speed which *combined* its linear and rotational velocities, while the underside of the ball, spinning away from the horizontal flight vector, would in comparison move more slowly. The sudden, drooping trajectories of forward-spinning balls, Newton reasoned, was owing to different air pressures exerted on the top and bottom sides of a ball; this imbalance "pushed" a ball downward.

Newton's analysis in this case (and so many others!) was largely correct—differential air pressures do indeed deflect the paths of spinning balls in one direction or another. But the Cambridge scientist's theory of the mechanics that govern the flight of spinning balls was simplistic. Newton's explanation did not take into account friction (or drag), which he disregarded.[2] Nor did he reckon with then unknown (and unsuspected) phenomena such as "boundary layer separation" or "wake turbulence." Another two centuries would pass until the theoretical physicist John William Strutt (Lord Rayleigh) published the definitive "On the irregular flight of a tennis ball" (1877). Yet even Rayleigh's careful analysis left a few mysteries that remain to this day. "There are two unsolved problems which interest me deeply," a contemporary physicist once joked; "[t]he first is the unified field theory; the second is why does a baseball curve? I believe that, in my lifetime, we may solve the first, but I despair of the second" (Adair 1990, 24).

Whenever an object moves through a fluid medium such as

water or air, it constantly shoves aside trillions and trillions of packets of hydrogen, oxygen, nitrogen, and carbon dioxide, along with much smaller numbers of other elements such as helium and carbon monoxide, and traces of things like mercury, lead, and strontium-90. If the object in motion happens to be highly streamlined (like, say, a bullet or a mackerel), the countless particles of fluid divide obligingly as the object passes through them and then come together again in smooth, flowing currents. Objects that are pointed at the front and have gently tapering sides are much better able to push their way through a fluid medium than things that are broad, cornered, or flat. This is why almost all fish evolved to have sleek, elongated bodies (the term for their typical, streamlined shape is "fusiform"), and this is why during the widespread fuel shortages that occurred during the 1970s, long-haul semitrailers were retrofitted virtually overnight with curved panels on top of their cabs. The rooftop fairings reduced the trucks' overall energy consumption by deflecting 70-mile-an-hour airflows smoothly past the cab and up and over the broad, flat faces of their cargo boxes.

In comparison with an 80-ton Peterbilt semi-truck, even one fitted with a full array of aerodynamic fairings, a baseball just out of the box looks smooth and slick. But aerodynamically the two objects aren't even close. A baseball, in contrast to a fish or a semi, is not a particularly streamlined shape. Even without considering those nubby red seams, a baseball (or any sphere) is aerodynamically rude; relative to its size, it's almost five times as hard to push through the air as the ungainly Volkswagen Beetle. Moving through the air at speeds of fifty to a hundred miles per hour, a baseball smashes into huge numbers of molecules of air that absorb some of its kinetic energy and impede its forward motion. This resistance is called drag, and, as you would expect, both drag and the energy necessary to overcome it increase with velocity. The faster you stir a thick Béchamel sauce, the more you must bear down on the spoon to keep it moving along. In the same way, a ball thrown at close to a hundred miles an hour must bludgeon aside twice as many molecules of air in any given amount of time as one thrown half that fast.

Of Magnus Effects and Boundary Layers

The forward spin of a curve ball causes its top side to smash harder against the stream of air molecules than its bottom side, exactly as Newton speculated. But actual tests to measure the forces involved were worked out first by Benjamin Robins (more on him in a later chapter) and subsequently by a German experimental chemist named Heinrich

Gustav Magnus. Born into wealth in 1802 in the state of Brandenburg, a short way north of Berlin, Magnus had come into the world as Joseph Meyer Magnus, the son of one of many European Jews (estimates run in the hundreds of thousands) who converted during that century to Christianity. Magnus's father's decision to renounce his ancestral name and faith was almost certainly more practical than ecumenical, as abandoning the faith of their fathers in favor of Christianity was at the time a necessary (if regrettable) first step for Prussian Jews who sought high social and professional standing.

The maneuver seems to have paid significant personal dividends for the former Immanuel Meyer and his family, and within a few years young Gustav was receiving private instruction from tutors in mathematics and the natural sciences. His training continued first as a student, subsequently as a professor of technology and physics at the University of Berlin, where he worked in a laboratory he himself had paid for using family money.

The youthful professor's expertise was in the relatively unglamorous field of mineral analysis. Magnus published his first paper in 1825 on the spontaneous combustion of iron, nickel, and cobalt, and shortly thereafter discovered the platinum-ammine compound with the jaw-testing name of tetraammineplatinum II/ tetrachloroplatinate II. (The name was soon abbreviated to "Magnus' Green Salt" in its discoverer's honor.) But Magnus soon became involved in broader pursuits. He shifted his attention from minerals first to inorganic chemistry, then to thermal expansion, magnetism, and ultimately straight-up, classical mechanics. He became especially intrigued by the varying movements of fluids over and around different objects, conducting experiments to measure the differences between free-streaming flows past objects that were stationary in comparison with objects that were rotating. It was this last area of research that led Magnus in 1852 to publish the scientific paper describing the principle of fluid mechanics that today bears his name: "Über eine abfallende Erscheinung bei rotierenden Körpern," or "On the falling-off tendency of rotating bodies" (Magnus 1853).

Magnus picked up the study of spinning objects just where Newton had left off. Like Newton, he observed that when rotating objects moved through a fluid medium, there appeared a mysterious, lateral force which deflected their movement away from a straight line. The spinning objects invariably deflected in the direction of their rotation, perpendicular to the axis of spin—just as Newton had written. Unlike Newton, however, Magnus began to study the phenomenon in his laboratory. The faster things spun, Magnus saw, the greater the amount of deflection caused by this new, vector quantity.

After conducting numerous experiments involving cylinders rotating in currents of flowing liquids, Magnus deduced that the amount of the deflection was dependent on two factors—the speed of the object's rotation (measured as angular velocity, ω), and the translational speed of the object (v) relative to the fluid. His calculations nailed down the math describing what Newton had observed on the tennis courts of Cambridge almost two hundred years earlier. (No matter that Magnus wasn't working with tennis balls in flight, but with rotating cylinders suspended in tanks filled with flowing streams of liquid. From an aerodynamic standpoint, it doesn't make any difference whether a spinning object itself moves through the fluid or the fluid moves past the spinning object.)

Magnus wasn't exactly sure *where* the force causing the deflection came from. But his experiments resulted in two important discoveries. First, Magnus showed that this novel, sideways vector was consistent and easily calculated according to a formula he himself developed: $Fm = S(\omega v)$, where F is the "Magnus force" and "S" is a measure of viscosity he included in the basic equation to permit comparisons between different fluids. More important, however, Magnus uncovered a dimension to the physics of spinning objects that Newton hadn't foreseen. Magnus observed that freely streaming fluids stuck *longer* to the surface of spinning cylinders and spheres that turned *with* the direction of flow than they stuck to the sides that happened to be facing into it. This phenomenon seemed both strange and significant, but Magnus could come up with no good explanation why fluids should cling to one side of a spinning cylinder more tightly than the other.

That riddle was solved some years later when another German physicist came up with the counterintuitive, through-the-looking-glass-physics of a weird place called the "boundary layer." Here's a simple account. Whenever any object—a baseball, say—speeds through the air, it must smash aside trillions and trillions of air molecules that get in its way. If you really could *see* the atoms themselves as they flow past a speeding baseball, however, you would immediately notice something odd, even shocking: none of the whizzing particles seem to be banging right against the ball's surface! Instead, close to the ball, something bizarre is happening. Or, not happening.

Clinging to the slick, white leather cover of the ball is a thin layer of molecules that seem not to be moving at all! These molecules are piled less than a millimeter high (about the thickness of a guitar string), and they're stuck to the front half of the baseball like a skin (Morrison 2013, 695). Even though the ball is rushing through the air at hurricane speeds, inside this shallow layer there's dead calm. In fact, if you were somehow shrunk small enough, while the ball was hurtling toward the plate you could chill on the MLB logo with a Margarita in one hand, iPhone in your other. ("A very, very small insect sitting on the moving ball," says Yale physics professor Robert Adair, "would feel no breeze at all" [Adair 1995, 6].) This serene space exists next to the surface of *any* object moving through *any* fluid medium. It's a strange, twilight zone first described in 1904 in a short, speculative paper by a young and (at that time) all-but-unknown German physicist named Ludwig Prandtl.

Not many people have ever heard of Ludwig Prandtl, and that's a shame. His light has remained, sadly, under a bushel. Even though Prandtl's discoveries were an important part of the great scientific discoveries of the early twentieth century, he was again and again passed over for a Nobel Prize. Prandtl's obscurity, as a few sympathetic biographers have suggested, may have been due to Nobel selection committees' open disdain for research that was being done in classical, mechanical physics. Pierre and Marie Curie had shared the Nobel prize in 1903 for their work in radiation, Lord Rayleigh won it in 1904 with the discovery of the element argon, and Philipp von Lenard won it in 1905 for his investigations into cathode rays.

Compared to mystery rays and hitherto unknown chemical elements, Prandtl's brilliant, mathematical explanations of what actually takes place when water streams around a rock—never mind that this very problem had baffled philosophers and scientists for thousands of years!—just didn't seem very sexy. Yet until Prandtl's exacting description of the physics of the boundary layer, no one—not one single person, ever—had explained accurately how fluids (they can be anything from water and diesel fuel to molten lava and air) flow around objects. It wasn't for lack of trying. Archimedes, Leonardo da Vinci, and Sir Isaac Newton had all tried—and had all failed—to figure out just how things move through wind and water.

The son of a professor of surveying and engineering at an agricultural college in Weihenstephan in southeastern Germany, Prandtl's entry into the world in February of 1875 was celebrated by his father (according to notes in his diary) at a shooting range "so that we could plant 12 conifers in memory of [his] birth" (Vogel-Prandtl 2014, 8). Young Ludwig grew up as a first and—as things were to turn out—precious only child. Two younger siblings died shortly after birth, a third was stillborn, and his mother's final two pregnancies miscarried. By the time he was a young lad, because of this terrible family misfortune, Prandtl's poor mother had retreated deep into the melancholy that lasted the rest of her life.

Growing up in what must have been a sad and lonely household,

Prandtl was free to pursue on his own the pastimes enjoyed by most young boys of that comparatively uncluttered era, spending long hours out of doors just to see how the world looked and worked. He roamed city streets, taking in the sights and sounds of urban life, peeking down alleys and crawling into storm drains to see where the underground tunnels led or to feel the vibrations of street cars rumbling overhead. Every day brought new surprises, new things to ponder. After one above-ground excursion, for example, Ludwig returned home to report that he had been amazed to see that women had legs "just like men" (his words) when wind gusts lifted their voluminous, multi-layered skirts.

The young Prandtl, according to his daughter and biographer, was especially fascinated to watch rain gutters fill with water and carry away leaves and bits of paper. Indeed, the sight of things flowing in streams seems to have intrigued Prandtl all his life. By the time he was twenty-six, he had obtained both his doctorate in physics and his mission in life, which was simply to figure out why fluids moving through pipes did not hug the walls (as one might expect) but flowed as a free stream down the middle. If this sounds like a mundane, perhaps even a trivial career path, the next time you take a ride in an airplane you might take a moment to thank Ludwig Prandtl for describing exactly how air flowing over its wings keeps you safely aloft with the birds.

Like many Eureka! moments in history, Prandtl's breakthrough scientific paper, "Motion of Fluids with Very Little Friction," is simple to understand—once you think of it (Prandtl 1904). His insight was twofold. Prandtl first assumed that as a fluid moves over and around an object, normal everyday friction would cause those molecules that touched the surface of the object to stick to it. The fluid didn't have to be a normally tacky substance like tar or molasses—he guessed that *any* fluid would likely cling to the surface of *any* object in essentially the same way. Prandtl deduced, therefore, that frictional forces at or near the surface of objects would have to be relatively great—much, *much* greater even than scientific geniuses like Archimedes and Newton had ever suspected—and that those forces would retard almost to a standstill whatever fluid happened to be flowing by. Those molecules that stuck to the surface of the object, in turn, would be bumped by more molecules that were attempting to stream by just above or beside them, and, because of myriad collisions, those molecules, too, would be slowed down or stopped altogether. It would be like being caught in a phantom traffic wave.

Yet this "no-slip" condition, Prandtl reasoned, couldn't extend very far away from the surface. Otherwise, beer would simply ooze from bartenders' taps and storm gutters would never be able to flush away debris.

Instead, Prandtl pictured a relatively thin *Grenzschicht*—a distinct "border zone" that existed very close to the surface of any object moving through any fluid. Within this "boundary layer" (as *Grenzschicht* came to be translated internationally), fluid molecules hardly moved at all. But farther away from the surface, fewer and fewer collisions would occur between stationary and freely streaming molecules, until at some clearly defined distance, virtually all parts of the fluid would be flowing essentially unobstructed.

Any fluid moving through or over *any* object, Prandtl concluded— it could be water in a pipe, say, or wind over a wing—thus had to be described mechanically with reference to *two* separate and characteristically distinct regions: a thin, surface layer where flow was minimal to nonexistent, and a separate region some slight distance away where flow was free, or (in the language of those who study fluid dynamics) "inviscid."

Prandtl's mechanics of the boundary layer proved to be essential in understanding and designing all kinds of things in the modern era from bullets to the wings of jet aircraft. But his theory also describes what happens when a rapidly moving baseball rams through the air. The molecules on the front side of the ball can flow smoothly over the boundary layer that clings to its curved, frontal surface.[3] Once the molecules pass beyond the circumference of the ball, however, the curvature of the surface is much less effective in directing their orderly, streaming flow, mainly because of all the temporarily open space that's constantly being created behind the ball as it shoves its way forward. Trailing behind the ball is a region of relatively low air pressure, and when the molecules forming the boundary layer encounter this less-dense air they tend suddenly to break free from the surface, swirling about wildly.

This molecular free-for-all is called "wake turbulence," and it exerts a significant backwards force (or drag,) on the moving ball, drastically retarding its momentum. As a curve ball spins its way to the plate, the boundary layer is dragged forward and down in the direction of rotation until it just passes around the bottom. At that point—about 7 or 8 o'clock on the ball, seen from the third base point of view—frictional forces are no longer sufficient to hold the molecules of the boundary layer on the surface, and they're flung rudely off, whipping violently rearward and up. (Picture yourself trying to free a car stuck in the mud by pushing on it from behind. When the wheel suddenly breaks traction, you'll get a face-full of goo slung up in your face.)

The faster the ball spins, the more that wake turbulence resulting from the separation of the boundary layer will be deflected upwards, and this angular vector is paired (as per Newton's third law) with a

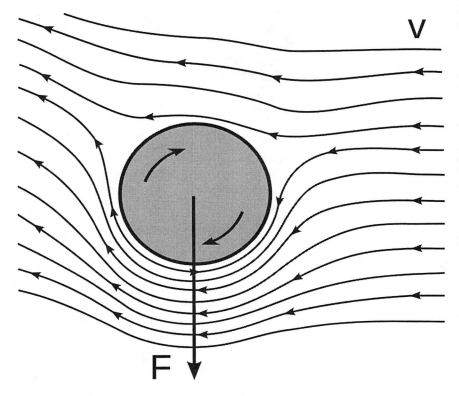

The Magnus Effect on a spinning ball. The ball travels from left to right, with topspin, creating a downward force dependent on translational and angular velocities (speed and spin) as well as on air density. The image shows also how the boundary layer clings longer to the side of the ball that spins in the direction of the airflow (here, the ball's bottom), deflecting wake turbulence upward and creating an additional downward force (Bartosz Kosiorek).

second, aerodynamic "Magnus force" that pushes the ball perpendicular to its trajectory. The spin of the ball pushes the wake upward, and the air pushes back down. Both forces combine to make topspin curve balls "bucket" and back-spinning fastballs "hop."[4] In the case of Sabathia's slider, which is spinning on a tilted axis, three vectors—gravity, boundary layer separation and wake turbulence, and the Magnus force—are concentrated toward the upper left quadrant of the ball, pushing and pulling it emphatically sideways as well as down. But really, a pitcher can make a baseball move to a greater or lesser degree in any direction he wants simply by changing the spin axis.

Some ball parks are located near sea level, others (like the Braves'

Truist Park in suburban Atlanta or the Rockies' Coors Field in Denver) significantly higher up where the air is less dense. Measure for measure, the air in these parks contains many fewer molecules that need to be shoved out of the way by objects moving through them, and so collectively they cannot exert as much influence on the trajectories of pitched or batted balls. This difference has real consequences both for pitchers and hitters.

Coors Field in Denver, for example, is known as a hitters' ballpark for two reason. The first and most obvious is that fly balls travel farther there because they must push aside fewer air molecules before reaching the outfield fences. In 2019, for example, batters at Coors Field hit about one more home run per game (2.7 against an overall MLB average of 1.8) than in other league ballparks. From a pitcher's point of view, that's bad enough. But this Rocky Mountain high handicaps pitchers in another, more significant way. Because the "air density index" at Coors Field is only two-thirds that of coastal Petco Park in San Diego (41.98 to 64.07, respectively, on a typical midsummer day), curve balls and sliders there don't curve and slide nearly so well as they do at sea level because there just aren't as many air molecules available for the physics of Prandtl and Magnus to take effect. "Pitching at Coors," says Jonathan Leshansky, "is like trying to defuse a bomb when you have a bad case of the shakes" (Schaffer 2007). Where to build the ideal ballpark from a pitcher's point of view? That would be in Death Valley's Badwater Basin, where "elevation" is a negative 282 feet and air density index on a cool winter's afternoon runs as high as 75!

When the Spinning Stops

With spin comes angular momentum, and with angular momentum comes stability. Anything that spins carries energy just because of its rotation; the faster a baseball spins around its axis, the more angular momentum it carries, and the less likely, therefore, it can be bumped off course by bats, cross-winds, or (as actually happened once) a meandering bird.[5] Since the kinetic energy of a rotating sphere is the product of its radius and its angular velocity, a fastball backspinning at 3,000 rpm carries twice the angular momentum of one spinning half that fast.

Depending on the orientation of the axis of rotation—top to bottom, side to side, or anything in between—different pitches will feel "heavier" or "lighter" to a catcher when they impact his mitt. A fastball backspinning merely at, say, 700 or 800 rpm will be much less able to resist the pull of gravity than one rotating three or four times as fast, causing it to fall more quickly; such a pitch, appropriately, is called a "sinker," and it's responsible for a lot of infield outs because batters who are anticipating a slightly less rapid drop will tend to swing slightly high, "topping" the baseball into the ground. Another common effect of a sinking fastball, however, is grouchy catchers with sore hands. An ex-college player described his experience catching sinkers: "[They] fucked my whole life up. Went through two gloves and ... fucked up my thumb on my catching hand since the ball kept hitting the bottom of the glove."

We can't leave the subject of spinning baseballs, however, without looking at the one pitch that doesn't spin at all. We've seen how control of a baseball's spin can be advantageous for pitchers who want to make things hard for the batter; now let's look at why the *lack* of spin, under the right conditions, might make a batter's life harder still. The pitch that lacks spin is called the knuckleball, named after the way its early practitioners gripped the ball when throwing it with fingertips bent inward.

Eddie Cicotte of the disgraced 1920 Chicago White Sox was nick-named "Knuckles" for his mastery of the pitch, but these days most pitchers who throw the knuckleball don't use their knuckles at all. They just dig their fingernails hard into the cowhide cover, so their finger-tips don't pull down on the ball and start it spinning when they let go of it. There's little or no wrist snap; the ball squirts out of the hand like a squeezed grape. Top pitchers in the majors these days throw fastballs that spin on average more than 2,500 rpm; their curve balls spin slightly more rapidly at around 3,100 rpm, or around 15 revolutions between release and home plate. In contrast, the best knuckler makes *less than half a rotation* over the same distance. Batters say the ball looks like it's "floating," "wiggling," even "dancing."

Pitchers who want to master the knuckleball must literally cast their fate to the wind. A spinless baseball in flight is influenced only by gravity, drag, and the fickle summer winds; except for gravity, these variables cannot be known in advance, and so a spinless baseball will be deflected in different ways with each different pitch.[6] Because it lacks spin, the knuckleball has virtually no angular momentum, and it tracks in flight much like fourteenth-century musket balls: that is to say, *nobody* knows where the damn thing will wind up. Former major league catcher Bob Uecker once revealed his strategy for dealing with a mean-dering knuckleball: "Wait till it stops rolling, then go pick it up." Much the same weird, wandering trajectories can be seen occasionally during soccer matches, too, where balls sometimes are kicked without spin. To describe their fluttering, erratic flight paths, Brazilians say *pompo sem asa*, a "dove without wings."

On June 26, 2019, I watched Game 3 of the College World Series, and I heard the following description of a pitch thrown by Mason Hickman, Vanderbilt's pitcher: "[It] spins straight top-to-bottom and holds plane all the way to the plate." The commentator's analysis of Hickman's fastball was spot on; I was both surprised and delighted to hear such useful and detailed analysis as part of televised baseball coverage.[7] But it was not the sort of thing one would have heard 50, 10, or even two years ago from a baseball broadcast booth, even from knowledgeable contemporary network analysts such as Ron Darling or John Smoltz, both MLB pitchers with long, successful careers. (Smoltz was elected to baseball's Hall of Fame in 2015.) Back then, Darling or Smoltz would have talked vaguely (though accurately) about the "hop" or "late life" on Hickman's fastball. But the technical language of spin mechanics and its effects on the trajectory of pitches just wasn't part of baseball lingo then, or at any time previously, for that matter. Now, at last, it is. The game of baseball is almost two hundred years old; finally, words have caught up to what pitchers have been throwing all along.

Buggy Whips and Guillotines

No matter what the game or what the ball used to play it, the same basic forces act on any spherical object spinning in a viscous medium such as air. Spin gives any ball stability *and* a measure of unpredictability. On the one hand, spin stabilizes a ball's overall trajectory; according to the conservation of angular momentum, the faster a ball spins, the more it will hold steady along its axis of rotation. On the other hand, any significant amount of spin causes the surface of the ball to interact with the airflow around it; the greater the ball's angular velocity, the greater amount it will be deflected in a direction perpendicular to its axis of rotation. The more rapidly a ball spins, in other words, the more it can be made to veer suddenly from the course an opponent's brain has plotted for it. The effect is particularly emphatic with a baseball because of its asymmetric construction, raised seams above a slick, smooth surface.

Pitchers use spin to make a baseball move in ways that seem suspiciously if not flagrantly otherworldly. But the national pastime is not the only sport where balls in play break, drop, bend, or, like those atoms described long ago by Lucretius, swerve mysteriously into the void. Similar "illusions" happen all the time in other ball games, too, all consistent with the laws of physics. Raphael Nadal hits with greater topspin than any other contemporary player (or any tennis player in history, for that matter). His lurching, kicking forehand comes off the racket spinning (at peak) at close to 5,000 revolutions per minute. The effects of the scorching spin Nadal imparts to a two-ounce tennis ball are tremendous; assuming two identical balls are struck with equal force, one hit with Nadal's aggressive topspin will strike the ground as much as *twenty feet* shorter than one that is hit without any spin, or "flat." Small wonder that to be able to hit tennis balls with spin gives a huge advantage to any player who knows how to use it.

"Big Bill" Tilden, the dominant player of the 1920s, wrote in his book, *Match Play and the Spin of the Ball*, that anyone who wished to play tennis at championship levels needed to learn how to impart spin to the ball. "Spin means control," Tilden advised those who wished to play competitive tennis; "never make any stroke without imparting a conscious, deliberate and intentional spin to the ball" (Tilden 1925, 2).

I played tennis all through high school, and by spring of my senior year I thought I was pretty good at it. I was playing weekly matches against several players then on the tennis team at nearby Albright College, an NCAA D3 (non-athletic scholarship) school, winning some and losing some, and once, at the posh suburban tennis club where I worked a seasonal job cutting lawns and grooming four sage, pulverized-rock courts, I played a no-stakes match against that year's finalist in the Pennsylvania state boys' scholastic tennis championships. The outcome of that contest was never in doubt—the final score was 6–1, 6–1—but I was able at least to avoid the shame of being bageled, and I came away glowing inwardly that I had stayed on the court with Bill Breitinger.

Twenty years later, when I stepped on another court to play against a serious topspin artist, a guy named Darryl, I was chagrined to see how the modern game had passed me by. This would have been in the mid–1980s, by which time the big, looping topspin drives first developed on the professional tour by Bjorn Borg and, later, Ivan Lendl were being imitated on amateur courts worldwide. Darryl regularly prowled the local high school tennis courts, looking for a game. He hit medium-paced forehands with a vicious, upward swipe.

The first time we played, he simply demolished me. If I won a single point that day, I can't remember it. Time and again I watched his forehands clear the net by several feet and swoop deep toward my backcourt; time and again, drawing on years' worth of experience gauging the probable trajectory of tennis balls, my brain confidently informed feet, legs, and arm that the ball was going to sail long and well out of court. Then, just as I turned aside to let the ball pass by, it clawed at the air, hit some heavy, invisible curtain, and bounced before my feet. The first time it happened, I was amazed. Next, after a few swings and mishits, frustration set in. How was I supposed to return a ball if I couldn't anticipate where it would land? Eventually, I was reduced to silly giggling. How in the world could he *do* that?

The topspin drive is the tennis player's version of a pitcher's curve ball. Take a closer look at Nadal's "buggy whip" forehand, for example. It's known by his opponents for its severe, precipitous drop and its cripplingly "heavy" feel on their racket strings when they try to return it. At peak spin, Nadal's forehand drive is rotating about twice as fast as that of his long-time opponent in Grand Slam finals, Roger Federer, and almost three times the rate of players from the previous generation like Andre Agassi and Pete Sampras. Anyone who thinks that the looping, high-bounding trajectory of Nadal's forehand is only an "illusion" of his opponent's brain can be swiftly enlightened by watching one of his matches on television where, one assumes, the camera's eye is unaffected by any physiological limitations peculiar to humans. The sudden, exaggerated plunge after Nadal's buggy-whip crosses over the net is as visible and as startling on a high-definition flat screen as it is to someone on the opposite side of the court attempting to hit it.

Because Nadal hits with such ferocious topspin, he can hit a ball as hard as possible and still count on Herrs Prandtl and Magnus to drop it well inside his opponent's court. And there's one more advantage to his heavy topspin. As soon as Nadal's looping, forward-spinning ball hits the ground, much of its kinetic energy is converted into a rebound; the ball strikes back up at opponents like a viper.

It's the other way round when a tennis ball is made to spin backwards. The tennis player's equivalent to a baseball that rises or "hops" is the floating underspin, or "slice," backhand. One of the best such backhands of all time belonged to Steffi Graf. Graf's powerful, flat forehand won points, but it was her "guillotine" backhand that kept her in them. Struck with a sharp, downward slice of the racket, the ball seemed preternaturally lazy; it appeared to be resting atop an invisible, magic cushion as it floated serenely from one side of the court to the other. Graf used the shot to control the pace of a ground-stroke rally; it bought her time to prepare for the next shot, and its low, skidding rebound was hard for opponents to return with power. In contrast to Nadal's high-kicking forehand, Graf's underspin backhand had almost no bounceback; as Chris Evert, Graf's long-time opponent and fan once described it, the ball simply nestled into the court and fell asleep.

Benders, Dimples, and
the Quarterback's Forward Pass

At least three other popular ball sports take advantage of spin. Soccer players, for example, rely on the same physics seen on tennis courts and baseball diamonds. *Bend It Like Beckham*, a film released in 2002, takes its title from the swooping, twisting flight path of soccer balls kicked by midfielder David Beckham. Beckham, a British football (i.e., soccer) player who began his career playing for Manchester United, became known for his ability to kick a soccer ball and set it spinning so that it would curve sharply (or "bend") around the goalkeeper and into the net. Beckham was hardly the first player to put spin on soccer balls—the legendary Brazilian star Pelé was doing the same thing in the 1960s and 1970s—but he more than anybody else made the technique famous.

The advantages derived from spinning a soccer ball are so well known, in fact, that players of any age or sex soon learn to do it. My granddaughters were taught how to "bend" the ball almost as soon as they stepped onto the field. Kick the ball slightly off-center and on the right side to spin it counterclockwise, but kick on the left side to spin it in the opposite direction. Either way, the rapidly spinning ball (kicked stoutly, it spins at about 700 rpm) will fly more-or-less straight for the first several yards. Then, as the mechanics described by Newton, Magnus, and Prandtl take effect, the ball will suddenly and sharply "bend" around and behind a phalanx of stunned defensive players.

Out on the links, hooks and slices are the golfer's equivalent of sliders and screwballs—the hook breaks left, the slice to the right. Valuable as these sideways swerves are in a ballpark or on a soccer pitch, however, neither is desirable for a golfer; one (or sometimes both) can be the bane of amateur players who habitually impart sidespin to their drives by unconsciously swinging across the ball instead of through it. A ball struck this way sails off the fairway into the weeds. And a golf ball that is accidentally "topped" when struck can be made to drop just as precipitously as any major league curve ball. In 2010, for example, a golfer who identified himself as "Matplusness" posted a desperate question on an internet golfing site called "The Sand Trap." His problem was consistently hitting over the upper half of the ball, thereby imparting ferocious topspin to it. "They get off the tee great for 30 yards," he lamented, "then take a nose-diving plummet to the depths of hell" (thesandtrap. com 2019).

Baseball pitchers are barred by the rules of the game from tampering with the balls they throw, though such proscriptions never stopped them from scuffing, scraping, or sliming a baseball to change its aerodynamic profile. A baseball with even a small imperfection on its cover will never move in the same way twice. You might then wonder: why are golf balls deliberately scarred with all those funny little pockmarks? It seems common sense that a ball with a smooth, unblemished surface ought to slip more easily through the air than one with nicks and scuff marks. In fact, the opposite is true. Early golfers discovered to their surprise that worn golf balls could be consistently struck *further* than brand new ones. Their discovery soon led sports manufacturers in the nineteenth century to stamp grooves and eventually shallow cups (or dimples) into the surface of their golf balls to make them stay airborne longer. Modern tests confirm that a ball pockmarked intentionally will travel about twice as far as one with a perfectly uniform surface.

The explanation for this oddity comes straight out of the laboratories of Gustave Magnus and Ludwig Prandtl. When early golfers hit a ball with nicks and scuffs on its cover, the imperfections significantly disrupted—or "turbulated," in the language of fluid mechanics—the normally calm air molecules in the boundary layer. One consequence of such increased commotion in the boundary layer was a slight increase in surface friction or drag. This of course tended to slow the ball down. But a surprising—and much more important consequence—was that when the boundary layer was "turbulated" it tended to cling for a *longer* time to the surface of the ball before it was torn away. The conclusion for the club sport was obvious: if you wanted to hit a golf ball as far as possible, it was better to exchange the *little* bit of additional drag caused by a few nicks and cuts for a *much greater* overall reduction in drag forces obtained by delaying the separation of the boundary layer.[8]

The addition of multiple indentations to the golf ball's surface—the first patent on such balls was issued in 1897—regularized this unintuitive aerodynamic principle before it was formally explained by Prandtl, and soon manufacturers began turning out balls in a multitude of different types and patterns of surface imperfections. The dimples—these days, typically there are between three hundred and five hundred of them per ball—turbulate the boundary layer, helping to keep it attached to the ball farther around its backside. And the pockmarks bring one other advantage: because the ball's textured surface allows the club head a better grip or "bite," golfers can impart to it a much greater rate of backspin. More backspin prolongs the ball's flight and stops it more quickly once it lands. Most dimples are hemispherical in shape, but the HX ball produced by Callaway covers the surface with an array of tiny hexagons, somewhat like a beehive. In any case the shape of the dimples seems to be less important than their depth, where variations of as little as one-thousandth of an inch can significantly change the ball's trajectory (Veilleux 2005).

The odd ball game out when it comes to making use of spin is football, mainly because a football is not really a ball at all but instead a "prolate spheroid." (This is the three-dimensional shape that you get from rotating an ellipse around its major, or longer, axis. The NFL game balls look the way they do because that's the shape of the original "pigskin" used in American football, a blown-up pig's bladder.) Yet spin has come to play an important a role in football matchups ever since September 5, 1906, when Bradbury Robinson of St. Louis University threw a twenty-yard forward pass to teammate Jack Schneider for a touchdown against Carroll College.

The "projectile pass," as it was then known, was relatively late in becoming accepted in American football, mainly because in contrast to the rest of the game, merely throwing the football didn't seem a sufficiently rugged way to move it down the field and into your opponents' territory. (Constrained by similar, nineteenth-century notions of "manliness," baseball players were at first reluctant to wear gloves to assist in catching balls. Even catchers, who obviously had the most to risk in terms of black eyes, smashed fingers, and lost teeth, didn't start wearing protective gear until the mid–1870s.)

To throw a tricky curve or slider you spin the ball rapidly enough to take advantage of the aerodynamic forces described by Newton, Magnus, and Prandtl. But to throw a football accurately enough to drop it in the hands of a running, twisting receiver forty or more yards away, you rely mostly on the conservation of angular momentum. Baseballs, soccer balls, tennis or volley balls are all perfect spheres and don't care particularly how you pick them up or which way you throw or hit them. Any old spin axis will work—top to bottom, side to side, even none at all if you want to throw a knuckleball.

Not so with a football, however. Just throwing it hard doesn't guarantee it will fly far, let alone true. If you can remember how a football flopped and fluttered the first time you tried to throw like an NFL quarterback, you know the problem. To throw a football forward, you must learn how to spin it side to side, or longitudinally. Here's how Russell Wilson of the Denver Broncos throws what's called a tight spiral. He grips the ball by placing several fingers of his throwing hand along the row of laces. These run lengthwise along one side of a football, almost like a zipper, and, because they are raised above the leather cover, the laces provide friction for Wilson's fingertips to pull against just as he lets go.

As with a curve ball, a sharp, downward pull against the laces sets the ball spinning around the ball's longitudinal axis. The spinning creates angular momentum, and with angular momentum comes directional stability. In the absence of any application of external force (a defender's tip, for example), the football will retain that initial longitudinal orientation for the entire time it's airborne. It's amazing that something so weirdly shaped as a football can fly stably for 50 yards or more. Without that tight spinning motion, however, those prolate spheroids would veer and tumble just as unpredictably as one of Phil Niekro's drunken knucklers.

Several other forces act on a forward pass. A Magnus effect, for example, curves the football right or left, depending on whether the quarterback who threw it is left- or right-handed. A right hander's throw spirals clockwise and so will bend slightly in that direction, but Hall of Fame passer Steve Young was a southpaw, so his passes veered left. This lateral, tailing movement of a longitudinally spinning projectile has long been known as "yaw" to gunners and artillery soldiers, but it's only recently been investigated for the way it affects the flight of footballs.

In an article published in both the *New York Times* and the *Los Angeles Times*, aerospace engineer William Rae from SUNY Buffalo describes the results of wind-tunnel tests he conducted on footballs. Rae theorizes that no fewer than three significant forces act on a forward pass, including aerodynamic torque, which pushes the nose of the football up; gyroscopic torque, or angular momentum; and the Magnus effect, which nudges the ball to one side or another. Whenever in his computer simulations of flight Rae leaves any one of these three forces out, the ball swerves seriously out of line. Will knowing the physics of the forward pass change anything about the game or the way it's played? Not likely, says Rae: "You won't see people change the design of the football or anything like that because of physics." Still, he says, "what we learn about the physics of football or basketball is fun for the rest of us, especially teachers" (Leary 1996).

CHAPTER 5

Big Wheels Keep on Turning

*When a fifteen-year-old girl climbs off her bike and climbs back on at twenty-five, it may seem only the ten year interval that her body has forgotten, so effortless is the return to mastery [that] someone from an earlier century or from a country without material objects might think—hearing the description of a girl gliding over the ground on round wings ... that it was an angel or a goddess that was being described.—*ELAINE SCARRY, *The Body in Pain*

(Re)Inventing the Wheel

A slim ladder stands next to a bold vertical line—the outer wall of a city, probably. On the rungs: five men, heavy-headed clubs tucked in their belts. The topmost figure, club in hand, braces one foot against the wall, preparing to surmount it. It's clearly an armed assault. The drawing—like much early Egyptian art—is blocky and flat. The scale seems child-like—figures too tall, wall too short. It's the ladder itself that's most improbable. It floats astride a pair of solid disk wheels connected by an axle, rolled into place by two men with pry bars (Köpp-Junk 2016, 1).

The artistic record of wheels in the pharaonic era begins with this image from the tomb of the prince Khaemhweset (twelfth century BCE) at Saqqara. It's the earliest picture of a wheeled device in ancient Egypt. Later, through the dynasties, images of wheels and wheeled vehicles will proliferate. Tombs, gravesites, newly uncovered interior walls of dwellings will show potters at their wheels, ox-drawn carts, and splendid wheeled chariots of the sort that must have been used by Ramses' soldiers when they drove Moses and the Israelites from Egypt.

Despite the importance of wheel technology in pharaonic times, however, we now know that the first wheels—these magical, "round wings," as Elaine Scarry so keenly portrays them—did not come out of

Africa. True wheeled devices seem first to have been used for transport in Bronze Age Europe, the Near East, and on the grassy steppes of the Caucasus. There is no written record of the people or cultures of that place and time, of course. Still, we can picture them because we know in considerable detail the language that they spoke: Proto-Indo-European (or PIE), the reconstructed mother tongue of the largest family of world languages ranging from Sanskrit, Greek, and Anglo-Saxon to modern Bengali, Italian, German, and English.

Whoever then lived in that region, says the linguist and cultural anthropologist David Anthony, must have been familiar with wheels and wagons because they used words of their own invention to talk about them (Anthony 2007, 36). A lexicon of PIE, according to Anthony, includes a root form, $k^{w}ék^{w}los$, which meant "wheel" or "wheels" but could also by extension refer to a circle, a cart, or any wheeled vehicle. The technologies of our forebears may have been primitive by our own standards, but one imagines them every inch as clever as we are when it came to using words; perhaps those early speakers, like us, might even have quipped "nice set of wheels" to refer to a kinsman's new wagon.[1]

Recently a team of archaeologists working at a site near Ljubljana in the small, Balkan country of Slovenia uncovered the remains of an ash wheel and matching oak axle that must have been rolling merrily over hill and dale at a time when the first pharaoh was still in his cradle. It's called the Ljubljana Marshes Wheel, and radiocarbon analysis proves it's approximately 5,150 years old (i.e., its manufacture dates from about 3100 BCE). There's no accompanying record of the wheel's maker or his community, but whoever built it must have had extensive prior exposure to sophisticated wheel technology. The disk was found next to a matching axle, proving that the assembly was clearly intended to be mounted under a two-wheeled cart that could have carried a load. Given the limited tools and technology of the time—the age of bronze tools was barely in its infancy—the Ljubljana wheel is an impressive piece of woodcraft. The maker assembled a solid disk made of two boards locked together by four separate, wedge-shaped tenons spaced regularly at right angles across the joint. An oak axle with a boxed end was fitted inside a matching square cutout in the center of the wheel, indicating that axle and wheel turned round together.

That the Ljubljana wheel was found next to an axle brings up a crucially important point. One sometimes hears talk about "reinventing the wheel" as if that act were the grandest accomplishment of human ingenuity. In truth, it isn't even close. To imagine a spinning, circular shape does not require a great leap of mind; from bubbles and daisies on up to the moon and the sun, nature is full of things that look round like

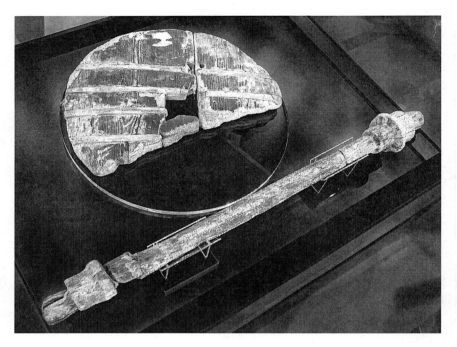

The Ljubljana Marshes Wheel is the oldest wheel and axle to be discovered. Uncovered in 2002 near the city of Ljubljana, Slovenia, the two-foot diameter wheel is estimated to be more than 5,000 years old. The squared end of the axle fits inside a corresponding hole in the center of the wheel, so both rotated together (Petar Milošević).

wheels. The real problem is figuring out what to do with a wheel once you have it. In 1913, Marcel Duchamp famously detached one wheel from a bicycle and mounted it atop a stool where it could be set slowly spinning. "To set the wheel turning," Duchamp said in an interview, "was very soothing, very comforting, a sort of opening onto awareness on other things than material life of every day. I enjoyed looking at it just as I enjoyed looking at the flames dancing in a fireplace."

L'art or *l'artifice*, Duchamp's installation laid bare a truth about the wheel: it's functionally useless unless it's mounted beneath a platform to support something one wishes to turn or to roll from one place to another. The intellectual challenge isn't to invent a wheel; it's to see a spinning wheel as one element of an unprecedented technology. "A wheeled vehicle," says Anthony, "required not just wheels but also an axle to hold the vehicle. The wheel, axle, and vehicle together made a complicated combination of load-bearing parts" (Anthony 2007, 65). Says professor of archaeology Heidi Köpp-Junk: "it's not the invention of the wheel that matters but rather the combination of wheel, axle, and transport-platform" (Köpp-Junk 2016, 44).

To conceive of that combination in the abstract is harder and more complicated than you might first think. Where would you turn for inspiration? "Thirty spokes meet in the hub," says the *Tao Te Ching*, "but the spaces between them is the essence of the wheel." Nothing in nature moves on wheels; they bear no resemblance whatsoever to other forms of locomotion like legs, wings, or fins. True, a few species of insects and small animals have learned the trick of rolling for purposes of defense or escape from predators. The ant-eating pangolin of Asia and sub-Saharan Africa tucks itself into a scaly ball to ward off predators, and the golden wheel spider of southern Africa flees from wasps by cartwheeling over dunes on the Namibian desert at speeds up to 2,500 rpm.

Impressive as these accomplishments may be when compared to human athletic potential, however, neither spider nor pangolin can in any way be described as "wheeled" creatures. Their primary mode of locomotion is with legs, and legs—as the designers of household or industrial "bots" are beginning to understand—are *much* more useful than wheels for getting around obstacles and over uneven terrain.[2]

The idea that wheels might move people or goods quickly from one place to another seems to have occurred relatively late among human cultures. If, indeed it occurred at all: in Olmec civilizations, for example (as throughout much of pre–Columbian America) wheels had indeed been "invented." But perhaps, as Jared Diamond has speculated, because of the lack of suitably tame, large draft animals, they were never thought to be more than curiosities or children's toys. There is specific and widespread archaeological and inscriptional evidence of wheels and wheeled vehicles *after* about 3400 BCE, both in Europe and the Near East. *Before* about 4000 BCE, on the other hand, any such references are nonexistent. The only explanation can be that prior to the beginning of the fourth millennium, there was nothing that looked like wheels and wheeled conveyances for humans to paint, to install in graves, or even to talk about.

Carts and wagons appeared near the beginnings of Bronze Age cultures spread throughout Eurasia. An image of a wagon on a clay cup is securely dated to 3500–3300 BCE, writes Anthony, and "the sight of wagons creaking and swaying across the grasslands amid herds of wooly sheep changed from a weirdly fascinating vision to a normal part of steppe life between about 3300 and 3100 BCE" (Anthony 2007, 300). The PIE speakers of the Bronze Age, according to Anthony, had at least four separate words they used in conjunction with wheels and wagons: two were words for "wheel," one for an "axle," and one referred to a thill, a thin shaft or pole attached to a wheeled cart to pull it. Since denominal-ization, or making nouns into verbs, is common practice with most languages, those same speakers, says Anthony, also used a verbal cognate to describe the act of travel by wheeled conveyance (Anthony 2007, 63). Perhaps those Bronze Age drovers "verbed" their trips with the same insouciance that Chamillionaire (Hakeem Temidayo Seriki) rapped about cops who saw him cruising twenty-first-century highways: "They see me rollin'."

Early wheels unearthed by archeologists are typically solid wooden disks constructed (like the Ljubljana wheel) of two or three heavy planks doweled together and cut into a circular shape. However, a pair of such wheels fixed directly to a rotating axle and placed, say, under a small, horse-drawn cart, will have restricted practical use. To move people or goods over serious distances, a wheeled vehicle needs to negotiate curves. This means that the wheel on the outer circumference of a turn must spin faster than its counterpart on the opposite side of the axle. Just as the tip of a whirling blade moves faster than a spot closer to the hub, the outer wheels of a wagon turning a corner must travel a greater distance than the inside wheels in the same amount of time. A vehicle with a pair of wheels fixed permanently to an axle can only move efficiently in a straight line; any attempt to turn it one way or another will cause one or both wheels to skid. Over the course of a journey, most of the time one of the wheels will drag rather than roll along the road.

The obvious solution is to permit inner and outer wheels to spin independently of one another. Attach a round, stationary axle to the underside of your wagon or cart. Then fit the outer ends of this axle inside matching circular holes located in the center of the wheels, enabling each wheel free to spin at whatever speeds the curve or road surface might dictate. This is indeed the kind of wheel-and-axle fitment most often found on artifacts in Kuban steppe gravesites; it's also the kind of arrangement used with the war chariots of Egyptian pharaohs and Mycenaean kings (Anthony 2007, 312).

Ingenious and simple as it may be, however, this technology creates a couple of problems. One of them is caused by the dirt that collects on the mating surfaces where wheel hub mates with axle. Even a paved road (and up until the twentieth century, there weren't many of them) is almost certain to be gritty, and without some way to keep the interface between stationary axle and spinning wheel hub free of dust, mud, and sand, the two adjacent surfaces will grind against each other like 50-weight abrasive paper, soon rendering the joint wobbly and weak. This kind of abrasive wear is troublesome enough. But a second problem—heat buildup between axle and wheel—is more intractable.

Whenever two surfaces in contact slide constantly against one another, they cause friction, and with friction inevitably comes heat. The greater the friction, the more heat is produced—enough heat, sometimes, even to set things smoldering.

Toward the end of the 1981 adventure film *Quest for Fire*, a young *Homo sapiens* woman teaches a company of Neanderthals how to create fire with the use of a primitive tool called a hand drill. This curious implement consists of a wooden spindle, rounded on one end, which is pressed tight against a matching, cup-shaped depression in a wood slab and spun rapidly with the hands back and forth until the heat generated by friction produces a glowing coal hot enough, with a little gentle fanning, to cause combustible material to burst into flame. Tools for producing fire in this way are thought to date back as far as 200,000 years, and the same friction-based, fire-making techniques are still taught to the Army's Special Forces and in numerous online tutorials for contemporary survivalists.

The physics behind this ancient device is simplicity itself: the kinetic energy of rotary motion with friction is converted immediately and efficiently into thermal energy. Similar principles apply to a wagon wheel spinning on a stationary axle. In effect, the combination behaves like a supersized hand drill; axle and hub start smoking and burn themselves up. Let's not lavish praise solely on whoever invented the wheel, therefore, or even on the person who thought up the wheel-and-axle. We should save some admiration for the genius who came up with a nifty little spinning device called a wheel bearing.

An Apology for Wheel Bearings

Friction between spinning wheels and the stationary axles that support them can be reduced by greasing the joint with some lubricating medium. But for people living before the petroleum age, finding substances that were plentiful, slippery, and long-lasting was no easy matter. Until the superior lubricating properties of petroleum derivatives were discovered in the late nineteenth century, the most widely used lubricant for wheels and axles was animal fat. Animal fats indeed possess excellent lubricating properties, but they proved only minimally useful for serious overland travel because their chief compositional ingredients, triglycerides, degrade quickly when subjected to heat.

Ancient philosophers and statesmen, when they weren't writing about history or politics, were pondering how best to grease the axles on their farm wagons. Among the substances recommended by elders

Cato and Pliny was boiled-down amurca, a bitter, watery sediment that collects at the bottom of vessels containing olive oil. (Cato also recommended amurca as the go-to solution for preserving figs, leather footwear, and all things made of bronze so they would not be attacked by verdigris.) Centuries of serious thinking on the subject brought no better solutions; teamsters in Renaissance Europe, for example, in their quest to keep their big wheels turning, stuffed their hubs not only with lard, tallow, and amurca but also with pitch, vegetable oil, even with a slippery paste made from crushed snails.

There's something perversely inspirational in using snails to help move a wheeled vehicle faster and further down the road. But snail smush, like all other slippery, animal-based lubricants, is incapable of withstanding the fierce heat generated by friction and the shocks and irregular vibrations caused by the inertia of a moving load carried on a wheel hub spinning on a stationary axle. What the transportation world *really* needed was a way to ensure that the contact surfaces of hub and axle move with respect to each other with as little sliding, or shear, stress as possible. One of the people who imagined new ways to reduce friction between rotating interfaces was the Renaissance polymath Leonardo da Vinci. Among the many papers collected and published after Leonardo's death were sketches for an intricate, Lazy-Susan contraption. His drawing shows eight wooden balls placed in individual compartments around a fixed hub and retained by an outer, circular band. A platform loaded with heavy weights rests on top of the balls.

Leonardo's drawings date from around the year 1500; his pictures and notes show how the circle of enclosed, freely spinning balls reduced friction by replacing continuous sliding contact with intermittent, rolling contact dispersed over numerous separate points. In his sketches, Leonardo shows the device laid horizontally, spinning on its side underneath a table on which it was then possible to shift great weights with relatively little effort. Leonardo's model would have worked in much the same way as the machines baggage handlers sometimes use to move boxes of cargo in and out of aircraft, sliding large, heavy containers back and forth and left and right on a grid of differentially spinning rollers.

More than a century later, Galileo Galilei described an almost identical spinning mechanism for reducing friction between a wheel and an axle. (Some say Galileo came up with the idea on his own. But others say that he had access to Leonardo's sketches and so merely commandeered somebody else's prior creation—as, indeed, he had done earlier with his famous telescope.) In any case, Galileo's model improved on Leonardo's in several ways. First, he substituted metal balls for the less durable wooden ones Leonardo had drawn. Even more importantly, Galileo put each ball inside its own ringed compartment (or "cage") and then looped the chain of balls and rings into a kind of circular bracelet.

A third generation of Leonardo's device did not appear for almost another two centuries, when in 1760 John Harrison, the brilliant British clockmaker, brought the concept of the "ball bearing" to perfection. Harrison wasn't particularly interested in wheels and axles; he just needed to make a timepiece sufficiently accurate that mariners on long voyages could rely on it keep track of the hour back at their home port. A clock that good, Harrison knew, would need gears that spun with as little friction as possible. To obtain such a high degree of time-keeping accuracy—his best timepiece kept time within a third of a second a day, a phenomenal achievement for that era—Harrison drew up the prototype for a spinning, caged ball bearing that was all but frictionless.

Modern wheel bearings, like the elementary particles of quantum physics, almost always come in matched pairs, positioned respectively on the inner and outer sides of a wheel where hub mates with the axle. A single bearing in turn consists of two steel rings or bands, each looking something like a cuff bracelet. Steel balls (or rollers) are loosely nested, or "caged" within the bands. The bands are called, respectively, the bearing's inner and outer "races," and everything is polished to such a degree as to seem inherently luminous. Grease—these days, it's not lard, olive lees, or a paste made of snails, but either a petroleum-based oil thickened with lithium, or a synthetic lubricant such as silicone—is packed into the interstices between balls, rollers, and races. The outer race is pressed tightly into the wheel hub, while the inner race encircles the axle shaft and spins freely around it.

Because balls or rollers and circular races touch one another only at random points of mutual tangency, surface friction between axle and hub is absolutely minimized. And because the balls (or rollers) are not attached to the inner and outer races that enclose them, they are free to spin simultaneously and independently of each other when the wheel itself turns, rotating backwards or forwards in relation to their races. This package of metallurgy, geometry, and physics is as fascinatingly beautiful as it's functionally ingenious.

I've watched when people who know nothing about auto mechanics pick up a brand-new wheel bearing. Inevitably they start goofing around with it; they spin it back and forth just to see and feel the different pieces move in their different speeds and orbits. Always, they grin; sometimes they giggle. No matter where they apply pressure—top, bottom, from the side—they simply can't stop the thing from spinning smoothly. More captivating than any fidget spinner, it's like playing with a miniature cosmos.

Times change. Interests shift. Everything flows. Once, I asked my teenage son what he thought the blues guitarist Blind Willie McTell meant when he sang, "Feel like a broke down engine, ain't got no drivin' wheel." Like most Millennials, my son had grown up with digitized screens and electronic devices, on the one hand, and a genuine appreciation for the natural world, on the other. For him, a summer weekend would combine playing with Game Boy while on an overnight camping excursion. But he lacked familiarity with basic mechanical devices, things that had formed an inevitable and essential part of my own growing up in the middle decades of the twentieth century. He had never seen and absorbed, for example, the complex system of knobs, gears, moving levers, belts, and spinning discs of a mid-century record player, and pictures of such engineered contrivances were not normally part of the constructions of his imagination. *Why would you feel like an engine?* he wondered. *What in the world was a driving wheel?*

He guessed, first, that McTell was talking about a car with a flat tire. Or maybe—then fourteen, he brightened as the possibility occurred to him—the singer meant the steering wheel? I tried to explain what McTell was talking about. It's not really anything specific, I said. McTell might be remembering an old-time steam locomotive he saw once in childhood before his eyesight was permanently lost. Or maybe he's picturing some industrial machine, a Depression-era tractor abandoned in some field. Whatever and wherever that engine is, it exists only in the memory of Blind Willie McTell, where it first appears, and then in the mind of the listener, where it resounds to become a mournful commentary on dreams, failure, and loss: *Feel like a broke-down engine, ain't got no drivin' wheel/ You all been down and lonesome, you know how a poor man feels.*

But it was no use. My son puzzled over the words of the song as with a joke that had to be explained, and the life went out of them in my telling. It was a misalignment of different generations I'd never experienced in quite the same way—amusing, yet unmistakably melancholy. Winter suddenly seemed a lot closer than it had in the days when I lay on my back in the mud, taking pieces off a broke-down Land Rover, believing that youth and strength would last forever.

Do Stagecoach Wheels Really Turn Backwards?

When my children were young, I tried several times to convince them that the wheels of stagecoaches in old cowboy movies never actually reversed as the driver pulled up beside the saloon. But to no avail:

my kids took it on faith that cameras never misrepresented reality—where was Jean Baudrillard when you needed him?—and they assumed that this brief, retrograde motion they were seeing was a universal characteristic of wheeled vehicular travel. Even as I tried to avoid laughing at their gross misperception, I was struck by the iron-hard reasonableness of children's minds. Doubtless, in their view of things, getting the wheels to turn backwards was a necessary prelude before coming to a stop.

We've all seen films in which spinning wheels seem mysteriously to reverse rotation, most often when a wagon riding on large, spoked wheels is pulling to a stop. First the wheels seem to be rolling forward along with the wagon, as you would expect; then, without warning, the spokes blur and start to spin *backwards*. The illusion persists for a second or two; then it blurs again, and wheels and spokes return to proper, forward rotation just as the coach comes to rest. Wagon wheels aren't the only things that act this way. Film sequences of propellers on aircraft show the same phenomenon—the individual blades turn one way as the engine coughs to life, briefly reverse themselves as they gain rotational speed, then reverse again, blur, and vanish. The impression that the wheels and propellers change direction *while they are spinning* is so strong it is completely convincing to children and adults alike. The world comes unhinged; what gives?

Propellers don't suddenly start spinning backwards, of course, and the wheels of stagecoaches and semi-trailers never turn in directions contrary to the motion of the rest of the vehicle. The illusion arises commonly in films, say researchers Dale Purves, Joseph Paydarfar, and Timothy Andrews, because there are always times in the filming of a moving vehicle when there is a difference between the speed of the wheels' rotation and the speed at which such scenes are filmed (Purves et al. 1996). When the stagecoach begins to move away from a standing start, for example, a camera typically will capture multiple images during the interval between any given pair of spokes, and the wheel will appear to be turning forward properly. As the spokes pass in front of the camera lens with ever-increasing speed, however, there will come a time when the intervals between revolving spokes and successive camera shots coincide. At that point, the wheel will appear magically to stand still. Further acceleration will again skew the relationship between spokes and camera lens' openings, but now for a brief time the camera will capture a view of the turning spokes slightly be*hind* the place they appeared in the previous shot, giving the distinct impression that the wheel has suddenly begun to spin backwards. The crazy illusion persists up to the point when the relative speeds of camera and wheel again fall in step; at this point the spokes once more appear to be turning in the direction of actual rotation, and they spin forward normally until they're going so fast they melt into a continuous blur.

Because the same illusion of wheels suddenly reversing direction can sometimes be seen in continuous light, however, some researchers say that the illusion—and it *is* an illusion—is at bottom a consequence of the way eye and brain process visual information. That crazy stage-coach wheel may be more than a camera trick, rather, an honest-to-God *trompe l'oeil*. Purves says that a similar phenomenon can be observed with spinning automobile wheel covers, airplane propellers, jet engines' fans, and other radially patterned objects *even when those objects are observed rotating in daylight with the naked eye.* This correspondence between the way rotating objects look in stroboscopic and in continuous light, Purves says, suggests that the different experiences may have some property in common. What might that property be?

The final word from researchers is not yet in, but it is possible, Purves writes, that eyes and brain process information not as a continuous stream of sensory data but as a series of discrete snapshots or what he calls "episodes." We see the world, he suggests, "by a series of fixations ... and [we] inspect pictures by fixations at about the same frequency" (Purves et al. 1996, 3696). You can even make spinning wheels do these crazy tricks on cue, should you want; all you do, reports W.A.H. Rushton in the journal *Nature*, is to hum a happy tune while you're watching. "Humming," Rushton says, "causes the eye to vibrate and this can produce a stroboscopic effect" (Rushton 1967, 1174). Suppose we really do perceive the world as a kind of perpetually un-spooling film-strip. Where, then, do you put your faith—in physics, or in your lying eyes?

Fairy Tales

One day in fourteenth-century Britain someone tells a story in which a young woman pricks her finger while preparing to spin flax. The tale takes hold, proliferates throughout Europe; Giambattista Basile, Charles Perrault, Jacob and Wilhelm Grimm all publish versions. Before "Sleeping Beauty" there was "Zirk Zirk," "Tarandandò," and "Rumpelstiltskin," and before "Rumpelstiltskin" there were "The Twelve Huntsmen," "Habitrot," "The Nixie of the Mill-Pond," "Mother Holle," "The Odds and Ends," and "The Golden Spinning Wheel." There were other tales of spinning wheels, too—tales like "East of the Sun, West of the Moon," "The Three Spinning Women," and "The Twelve Huntsmen."

And still more: "Spindle, Shuttle, and Needle," "The Lazy Spinning Woman," "The Girl Who Could Spin Gold from Clay and Long Straw," and "Twelve Brothers," a German folktale in which the heroine shuns all society for seven long years just so she can spin cloaks of nettles to disenchant her beloved brothers.

Imagine life when all clothing was made by hand. Imagine, for example, living in London in 1332, when the Subsidy Act allowed you to own tax-free just one dress, one pair of pants. If you were a peasant, you wore one set of clothing, rarely washed, made by yourself (if you were a woman), by your wife (if you were a man) or mother (if you were a child). Everything you wore was made by hand from wool or linen; before and up through the Middle Ages, every cap, robe, and tunic, every single bolt of cloth or skein of yarn, was begun by twisting clumps of loose fibers into long, continuous strings. The color palette for clothes was just as limited as their fabric—monotone browns, grays, now and then for variety or special circumstances, something resembling red.

Or picture daily life in colonial America. The buzz of a foot spinning wheel, the low-pitched humming of the "great wheel" or "walking wheel," supplied a constant, background music in your household. The familiar, droning sound could he heard even from a distance as visitors approached your village stoop or drew nigh to your cottage door. Lyndon Freeman of Sturgis, Massachusetts, recalled the "soothing music" of his mother's flax wheel that lulled him to sleep each night of his childhood.

Before there were tennis balls and gyroscopes, before there were ballerinas and wagon wheels, even before there were potters' wheels, butter churns, and screw pumps, women were spinning yarn out of wool, order out of chaos. I talk with Jen Steele who's become something of an expert on the subject because she not only spins the yarn, but she also grows the animals it comes from. Steele lives twenty-five miles from the nearest town on a ten-acre off-grid homestead in North Idaho. She's lived there with her family since 1997, stitching together an economy modeled on Thoreau's home-made basket—good enough for personal use, but not fancy enough for selling. On the day I talked with her she told me she had been spinning for so many years she could no longer recall when exactly she'd taken it up. "Lots of time my purpose is just to relax," she said. "But that's because we live in a day and age when we can go to Wal-Mart and buy a shirt. We can do these crafts because we want to."

Around us as we talk are trees, a handful of domestic animals, the familiar clutter of tools and smells and half-finished projects that characterize any active, rural homestead. A black-and-white border collie nudges a stick against my leg, trying to initiate a game of toss and fetch. A rooster crows; kids squabble just inside the house, briefly drawing their mother's attention. And all the while beside me I hear the soft, steady whispering of Steele's spinning wheel. I watch as she adds in more bits of wool, constantly twisting any loose fibers together, adding to the ragged end of the thread until it has grown to a couple of feet long. "There's so much lore about spinning," she says. "Wool, nettles, bison hair, fibers from a bush, all spun on a stick. You can literally spin yarn on a stick."

Steele turns from the machine, gently plucks a lock of carded wool from a small bag. To begin the process the spinster smooths the wool so the individual fibers all run in the same direction. The motion is basically no different from combing your hair. She wraps the lower end several times round a small, hand-held spindle, then gives the tool a firm twirl and drops it. (That's why it's called a "drop" spindle.) It dangles down, still spinning, spooling the length of thread she's just created. The spindle Steele uses had been manufactured recently, but it is in no essential way different from those that would have been used in prehistoric times, perhaps to spin yarn for hoods from sheep's wool or threads to make tunics from fibers derived from nettles or flax. It's a tedious, two-handed operation—clumsy the first time you try it, as with many basic manual tasks, but simple and efficient once it's mastered. The spindle is just a short, thin stick, a handy place round which to wind a thread.

It may well be impossible, as Ecclesiastes says, to make straight that which God has made crooked. A trick just as good, however, is to turn something short into something a lot longer. It's not known when or where it was first discovered how to make short animal hairs or plant fibers into long, continuous cordage. The art of spinning—"dynamic gluing," as the biologist Steven Vogel once called it—is at least 10,000 years old, and with its invention, humans were freed from their historic and limiting dependency on animal skins for clothing (Vogel 2016, 207). Steele pinches more fibers and twiddles them back and forth between thumb and fingers. This deft gesture twists the mass of loose strands into the beginnings of a single, continuous length of thread. The individual fibers are quite short, of course, a little shorter than they grew on the animal. Any single fiber is weak, also, no stronger than a human hair. They're somewhat stronger together, of course. But if you also *twist* them when combining them, something extraordinary happens. Because the fibers are no longer parallel, now when you pull on them, they jam against each other in a crosswise direction, securing themselves by shear stress, or friction. "The whole point," says Steele, "is to add twist to the fibers to lock them together."

Steele's handwork is as old as time: pinch a bit of wool, twiddle loose ends together, give the spindle a tweak to keep it turning round, spin it into yarn. It helps if you attach a small disk to the spindle to increase its angular momentum. The extra weight of the whorl (as this clever, supplemental device is called) makes the spindle rotate both smoother and longer so the spinner—or, later, "spinster"—doesn't have to tweak it nearly as often. Whorls come in different sizes—you would use a relatively light whorl to begin with, then load heavier ones as the weight of the thread and spindle grew. A broad, light whorl spins for a longer time, while whorls smaller in diameter (remember the twirling skater?) spin faster. While the spindle turns, the newly created (or "spun") thread wraps round and round it until it's covered with a thick, carrot-shaped batch of yarn. "Slower by the hour," Steele says, "faster by day." That's the difference, she says, between creating yarn on a drop spindle and on a machine. Over the course of any given hour, a machine makes yarn much faster. But you can't stick a spinning wheel in your apron pocket or handbag and create thread whenever you have a few spare minutes as you go about the rest of your day.

I'm saddened by the tedium of the work. What must it have been like, I wonder, to be a girl of twelve or thirteen and to see this life yawning to swallow you? "It would get real old real fast," says Steele, "if you had to do it for a living." Young Ellen Rollins of New Hampshire, who grew up in the early nineteenth century, wrote that "the moaning of the big wheel was the saddest sound of my childhood. It was like a low wail," she said, "from out of the lengthened monotony of the spinner's life" (Larkin 1988, 26). Small wonder that a suspiciously large number of fairy tales feature young women who hate to spin or, more often, who are punished by being made to do it.

One such story is commonly titled "The Three Spinning Women." "There was a girl," as the version of the story collected by the brothers Grimm begins, "who was lazy and would not spin." Angered by her disobedience, the girl's mother tells the queen (who just happens at that moment to be passing by) that her daughter loves to spin. Alas, she cannot pursue the work she longs for because the family is too poor to buy her flax. So the queen offers to board the girl at the palace where she will be given all the flax she needs; as a bonus for her industry, she is promised in marriage to the queen's eldest son. The unhappy girl does not know what to do, for though she loves the prince she knows also that she cannot bring herself to spin flax, even if she should live three hundred years.

By chance, however, while the girl sits in despair by a window, the palace is visited by three strange, old women who have spent their lives toiling at spinning wheels. One woman's feet have grown broad from constant peddling; the lips of the second are fallen grotesquely after years spent licking, and the third has a thumb worn big and flat as a boat paddle from twisting thread. Seeing what lies in store for his wife, the prince immediately absolves the girl from any hateful spinning of flax, after which she presumably lived as contentedly and free from drudgery as if she could shop for clothes from Stitch Fix.

I picture women stuck in garrets, fingers callused, minds spun away during hours, days, years. Still, that's not the whole story. The more I watch Steele work, the more I begin to marvel at her dexterousness. The movement of the spinner's hands and fingers is mesmerizing. The act is supremely graceful; it seems loving, almost reverent. Her gestures mix simplicity and mystery, science and art. Small wonder that spinning and spindles are integral to mythology and folklore throughout the world. It's above all a creative act, making something out of nothing.

Some days later, I talk with Alayna Ferran of Moscow, Idaho. Ferran has been taking classes in spinning at a store called Yarn Underground; the name of the shop, located downtown, plays off the counterculture movement from half a century ago and also documents the actual location of the store halfway below sidewalk level on South Jefferson Street. In making her own yarns and fabrics, Ferran isn't trying to save money; at 24, she's never known a time when clothes weren't amazingly cheap. She sees spinning from a different perspective. "It's satisfying, in a way," she says. "I'm taking something that's a big mess, something that's weak on its own, and turning it into something neater, stronger." Sitting at her wheel, Ferran also finds society in the lonely work of spinning, communing empathically with countless women before her: "I think, wow, this is something my great-great grandma did. I know she had a wheel."

In the final book of the *Republic*, Plato describes wandering souls who glimpse the structure of the universe and see for the first time "the spindle of Necessity, by which all the revolving spheres are turned." Ovid tells the story of Arachne, the daughter of Idmon, who bested Athena in a weaving contest and was turned into a spider. The Teutonic goddess Hulda was celebrated in pre–Christian folklore for the magic of her spinning and weaving; during the winter months, according to legend, she goes from house to house to see if the women and children are all spinning diligently (Peuchert 1993, 263).

At some point, a brain made the connection between the spinning of loose fibers into threads and the massing of random events into the shape of a life. The act of spinning loose fibers into cloth crossed over into the spinning of random events into stories, or "yarns." The connection between the two is almost as old as language. "Every spinner," says Rebecca Solnit, "takes the amorphous mass before her and makes thread appear, from which comes the stuff that contains the world, from a fishing net to a nightgown. She makes form out of formlessness, continuity out of fragments, narrative and meaning out of scattered incidents, for the storyteller is also a spinner or weaver" (Solnit 2014, 131).

Of Bicycles and Ferris Wheels

One of the seasonal pastimes for children in the small Pennsylvania Dutch town of Wernersville in the 1950s was to congregate each September in a vacant, grassy block between Gaul and Washington streets for the town fair. The plot was unoccupied and uninteresting during all the rest of the year, but for that one week in fall it might have been Oz; in the mind of a boy, the nighttime array of carnival attractions was as strange and beguiling as any munchkins and talking trees. Strolling the pathways between booths, one could win cheap prizes by shooting at a moving parade of cutout ducks with real, small caliber rifles or by chucking darts at a wall of balloons. And once, I remember, I came home with five neon-dyed chicks I had won by tossing ping pong balls into fishbowls.

At the western end of the plot, running alongside Stitzer Avenue, were the rides. There was a carousel popular with pre-school children, a small open-car train, go-karts that could be steered only along preformed rails, and one or two other machines that were slightly more intimidating; I recall a tilt-a-whirl ringed with naked light bulbs, a few always burned out, that midway through a session rose almost to vertical. Rising high above them all like the prow of a ship was the Ferris wheel.

His name was Wilson Thompson, and at the time I knew him he was the town paperboy, responsible for delivering to subscribers the *Reading Times* in the morning or, in mid-afternoon, the *Reading Eagle*. Thompson was afflicted from birth with cerebral palsy; he walked with a lurching, rolling gait, talking or humming softly to himself. He lived on Stitzer Avenue with his parents, a few doors down the street from my father's cousin, Luther Brossman. An elder brother, Army 2nd Lieutenant John Thompson, I was told, had died during World War II; that the brothers' fates should be so different and unfair, I remember pondering at the time for possible clues with respect to divine intentionality.

Day in and day out, Wilson Thompson rode his bicycle along the back streets of Wernersville, flinging papers up walkways or onto porches. But each fall he took one week off to ride the ferris wheel. He bought hundreds and hundreds of tickets, spinning on the wheel for hours on end. One week later, just before the fair closed, its sponsors raffled off a new Schwinn bicycle. With odds that must have been substantially better than even money, Thompson won the ticket draw. He did the same thing the next year, and the year after that, too, trading a week of rides on the Ferris wheel for a new bicycle for his paper route, and then the 50s were over and I grew up and out of amusement parks and lost track of Wilson Thompson and his quirky, instructive ways.

George Washington Gale Ferris unveiled the carnival wheel that would bear his name because he wanted something to display at Chicago's World Columbian Fair in 1893. That fair was intended to commemorate the 400th anniversary of Columbus's "discovery" of North America, and its sponsors commissioned the building of something that would rival the Eiffel Tower for grandeur and iconicity. Ferris, by his account, sketched out his giant, revolving passenger wheel during a meal at a Chicago chophouse. There's little comparison between that first wheel and smaller machine that so enchanted Wilson Thompson of Wernersville. Ferris's creation towered over anything nearby. (And further away, for that matter: at 264 feet tall, George Washington Gale Ferris's revolving wheel was at the time taller than any city building.)

Ferris's wheel was a simple adaptation of ancient, bucketed wheels built for raising water up from streams. At the time of their invention, these hydraulic noria (from Arabic *naura*, classical Syriac *na'orata*, "water wheel" or "growler") were understood to be a gift from the gods. "Cease from grinding, ye women who toil at the mill," proclaimed Antipater of Thessalonica in 85 BCE, "for Demeter has ordered the water nymphs to perform the work of your hands, and they, leaping down on top of the wheel, turn its axle which with its revolving spokes, turns the heavy millstone."

The first such water wheels may have appeared in Egypt in the fourth century BCE, possibly at about the same time in India; similar, contemporary devices are still in use in the Middle East and Asia in much the same way as they were used in antiquity. From early on, these wheels might have served pleasure as well as commerce: "It is easy," says historian Norman Anderson, "to imagine fun-loving children grabbing hold of a bucket for a free ride to the top and back down. Of course, it would have been a wet ride, but on a hot day this would have added to the pleasure" (Anderson 1992, 3).

Anderson ought to know what he's talking about; he's the author of *Ferris Wheels: An Illustrated History*, a comprehensive history of passenger-carrying wheels built expressly for amusement. Large, bucketed "whirligigs" or "ups-and-downs" (as such wheels were called) were common sights in Renaissance England at Bartholomew Fair at Smithfield where, according to one chronicler, "children in flying coaches ... who insensibly climbed upwards, knowing not whither they were going, but being once elevated to a certain height, they came down again according to the circular motion of the sphere they moved in" (Anderson 1992, 7).

Noria were repurposed in the nineteenth century along riverbanks in the Pacific Northwest and in Alaska where commercial fishermen and native Americans alike used the churning wood buckets to lift migrating salmon out of the water, then dropping the disoriented, squirming fish rudely into a catch basin. Fish wheels proved so efficient at plucking migrating salmon from the Columbia River that they were subsequently banned to save the species from extinction. Lately, however, in an ironic reversal of the applications of technology, the wheels have been resurrected not to harvest fish for canning but to let conservation officers catch and count them, tag them, and then release them back to the river to continue their final journey upstream.

Salmon presumably dislike being flung around in the buckets of a turning wheel, but humans ride these big wheels just for the pleasure of a taking a spin. Large, passenger-carrying wheels have long been sources of novelty and wonder; in centuries when flight was an absurd dream, a tall pleasure wheel, much like a hot air balloon, offered curious townspeople the only way to get a bird's-eye view of the local landscape.[3] The view from the topmost bucket of a gigantic whirligig gave an elevated, God-like perspective on the world. Looking down and out was a stimulus for wonder, for philosophizing: it's from a position in the carriage of Vienna's giant "travel wheel" (the Wiener *Riesenrad*) that a lead character in *The Third Man* (1949), a British *film noir*, delivers his memorable, cynical summary of civilization:

> In Italy, for thirty under the Borgias, they had warfare, terror, murder, and bloodshed, but they produced Michelangelo, Leonardo da Vinci, and the Renaissance. In Switzerland, they had brotherly love, they had five hundred years of democracy and peace—and what did that produce? The cuckoo clock.

If George Washington Gale Ferris didn't invent the wheel that bears his name, however, he surely built one to a scale that dwarfed any wheels that had come before his. When the young engineer sketched out a preliminary design during a dinner with some friends, he was already contemplating (his words) "a monster" (Anderson 1992, 43). Ferris envisioned a tension machine constructed along the lines of a bicycle wheel; it would have a diameter of 250 feet (scaled back from an original 300 feet), and it would—or so he dreamed—be capable of spinning as many as a thousand people at a time. Construction began in the middle of the winter of 1893; by the end of March the twin supporting towers were in position, the axle installed a few days later. (Transported to the build site atop a string of railroad flatcars, Ferris's axle was at the time largest single piece of steel that had ever been forged.)

The first paying passengers climbed into the buckets on the morning of June 21, 1893; by the time the Fair shut down more than a year later, more than a million and a half patrons had spent fifty cents each (about $15 in today's economy) to take a skyward spin—almost one in ten of all the people who had come to the fair. The wheel was dismantled in 1894 but soon reassembled the next year near Lincoln Park where for ten more years it took passengers for fifty cents a ride. Then once more it was taken down and moved, this time all the way to the banks of the Missouri River for the St. Louis Fair in 1904 commemorating the centennial year of President Thomas Jefferson's Louisiana Purchase.

Forget about modern amusement park rides like Avatar Flight of Passage, Cobra's Curse, or Disney's Tower of Terror; Ferris's wheel was the first, and if its patrons' testimonies are to be believed, it's still among the best. One of the most memorable descriptions of what it was like to take a spin on the monster comes from William Gronau, one of the engineers who supervised the wheel's design and construction. As one who knew the machine from inside out, so to speak, Gronau might be expected to give a rational account of his experience. Yet even the man who built it seems to have been blown away when he first took in the view from on high. "The sight [was] so inspiring that all conversation stopped," he remembered. Then he added: "The equal of it I have never seen, and I doubt very much if I shall again" (Anderson 1992, 62).

Flywheels, Prayer Wheels, and the Execution of St. Catherine of Alexandria

Not all ancient wheels were used for warfare, transportation, agriculture, or summer afternoon sport. It's easy to overlook the abundance and spectacular ingenuity of the rotational devices invented by humans over the last several thousand years. Among the earliest wheels were round, horizontal disks used to facilitate pottery making. A watched pot never boils, and a square pot (like a square egg) would be inherently weak at the corners and unsuitable for keeping its contents safe. A circle, on the other hand, is an ideal shape for a container because it resists being deformed under pressure both from inside and out.

It's possible to approximate roundness in making a pot by piling numerous, thin coils of clay atop one another on a flat, circular base, rather like stacking a rope or garden hose, and some early containers were constructed in just this fashion. But pots and bowls made in this way tended to be crude, more often "roundish" than truly cylindrical. It's easier and better to shape a vessel round if you spin it while it's under construction. Once the mass of clay begins to be whirled about, it will start to stretch outward, just like a blob of dough tossed into the air by a skilled pizza chef. Then, if you hold your moistened fingers (or some kind of flat-bladed carving tool) steady against the rotating, pliable clay, tangential acceleration (which you feel in your fingers or through the tool as "centrifugal force") will help you make the pot perfectly round.

The first and simplest mechanism used to spin wet clay into a pot or bowl was a lightweight, flat surface, something like a woven mat or (in the Middle East) a palm frond that could be turned so that the worker could shape first one side, then others. A much better solution, however, was to set the pot on a platform called a tourney, or "slow wheel." This was a portable, circular stone centered on some kind of pivot or axle. Instead of coiling the strips of clay atop one another by hand, one end of the strip was positioned on the stone atop a soft clay base. As the turntable was rotated, clay rope was gradually fed onto the coil until the pile reached a desired height. Further turning allowed the crude vessel to be spun slowly while hands and fingers shaped it into a perfectly round and smooth vessel.

The earliest potters' wheels (c. 5th millennium BCE) were slow, cumbersome tools, and their use was not widespread. But toward the end of the third millennium, a bigger, heavier, and smooth turning "fast wheel" came into use. Once it was attached to a vertical axle and mounted on a polished granite bearing surface under a worktable, a potter could shove this circular, heavy stone vigorously a few times with his foot (hence the common term for the apparatus, "kick wheel") at which point it would have built up enough angular momentum to keep spinning without any further attention for a usefully long time.

Ever since the Industrial Revolution, this kind of heavy, spinning disk has commonly been called a "flywheel." A flywheel is simply a relatively heavy object, usually a wheel or solid cylinder that can be set spinning on a central shaft. As soon as it begins to rotate, a flywheel absorbs most of the kinetic energy used to set it in motion and stores it as angular momentum. As with any spinning object, the quantity of angular momentum depends on its mass and angular velocity; the heavier the flywheel and the faster it rotates, the greater the momentum it carries and "conserves." This makes a heavy, spinning wheel a handy way to store energy temporarily in one place, gradually dispensing it as needed.

A potter in Ur at the beginning of the third millennium BCE, for example, could kick a 40-pound wheel stone (or *tournette*) a couple of times to start it turning atop its black granite base. A nub protruding from the bottom of the upper, pivot stone fit snugly inside a matched, polished socket in the lower stone. Some lubricating substance—animal fat or vegetable oil, probably—would be applied to the mating surfaces between nub and socket (Powell 2016). Once his wheel was in motion, the potter could concentrate exclusively on shaping soft, wet clay into a perfectly circular bowl or vase for as long as a half-minute before he needed to kick his stone once again.

As steam- and gasoline-powered engines became ubiquitous over the last few centuries, flywheels have had one extremely important application. Depending on the design of the engine in your gasoline- or diesel-powered vehicle, anywhere from four, six, or eight separate power impulses occur during every complete revolution of the crankshaft, thousands of times a minute, each one smashing about ten tons of force straight down. Such fearsome, interruptive hammering would quickly destroy any engine, even one with the finesse of the Ferrari Colombo, a twelve-cylinder design legendary for its inherent smoothness.

This intermittent power delivery is a design flaw inherent in any piston engine. For this reason, there's a thirty-pound chunk of metal (usually iron or steel) bolted to the back of the engine's crankshaft where its continuous rotational energy transforms the discontinuous impulses of individual pistons into a more-or-less continuous supply of power. As the flywheel spins with the crankshaft, it absorbs the energy of each separate piston power stroke, stores it, and carries it forward, in the process supplying energy to shove the next piston in the firing sequence up the cylinder bore on its compression stroke.

If flywheels can be used to smooth out irregular pulses of energy, they can also be used to keep it available long term on standby, so to speak. This property too is a consequence of the laws of inertia. So long as a flywheel keeps spinning, it retains its angular momentum—as much as 90 percent of it, if the bearings on which it rotates are relatively frictionless—which can then be retrieved and put immediately to use. In 2011, Beacon Power began operating the nation's largest flywheel-based electricity storage plant in Stephentown, New York, where 200 separate flywheels, each weighing more than a ton and spinning at 16,000 rpm, keep on tap the kinetic energy to generate up to 20 megawatts of electric power, enough to supply about one-tenth of the electric needs for the whole state.

Conventionally designed flywheels made of iron or stone and spinning on steel bearings would dissipate a significant amount of their stored energy in heat and friction, but newer flywheels are constructed of composite materials, and they spin on a virtual axis suspended within a magnetic field inside a vacuum chamber. They are outfitted with magnetic bearings (these "bearing surfaces" don't bear on each other at all!) and can store and subsequently return nearly all the energy put into them (Harder 2011). Such flywheels are likely to play increasingly important roles in industry as well as everyday household life. As one final advantage, says physicist Helen Czerski, flywheels come in all sizes: "a small one to go with the solar panels on your roof, or a vast bank of them to moderate spikes in the whole energy grid.... [They] are even being tried out on hybrid buses, storing energy as the bus brakes and supplying it back to the wheels when the bus needs to speed up again" (Czerski 2016, 194).

You may be wondering why a ponderous, spinning disk is called a "fly" wheel? The name seems absurd for something neither airborne nor insect-like. Figuring out where the name "flywheel" came from took a good bit of sleuthing. For once, the Oxford English Dictionary was little help. The first recorded use of the word came in 1784 near the beginning of the Industrial Revolution in Europe; people then began to speak of "flywheels" along with *stovepipes*, *tea caddies*, and *the reign of terror*. Searching for truth on Quora—always a hit-or-miss business—I found one self-proclaimed expert who asserted that the name "flywheel" conjured up a wheel that turned very rapidly, as in the expression "let it fly!" A writer on another website posted confidently that the name "flywheel" referred to the wheel's position "floating in space." Clearly all this was online nonsense. These explanations shed no more light on the subject than the old Marx Brothers radio show named for a shady law firm called *Flywheel, Shyster, and Flywheel*.

I tried a flanking approach to etymology: the German word for flywheel is *Schwungrad*, or "inertia wheel." This seemed more logically expressive than "a wheel that flies," but it didn't help me to construct a link between the English word and the function of the object it referred to. At last, I came upon a sketch of something called a screw press. This was a mechanical device used by ancient Romans to make wine and olive oil. The implement looked like a very large corkscrew; as workers spun a pair of wooden handles, a spiral, worm gear mashed the grapes or olives flatter and flatter, squeezing out the juice.

A similar, much more robust such machine was adapted to a different purpose in England after the restoration of the monarchy in 1660, when a large screw press was used to strike coins with an image of Charles II on them. (Charles I was decapitated in 1648, and his head soon vanished from all coins of the realm!) When a replacement Charles came to the throne, a complete set of coins featuring the new royal head had to be minted right away. Tens of thousands of crowns, half-angels, shillings and ha'pennies were punched from sheets of metal by workers who spun the horizontal levers of a large screw press, forcing the cutting die down onto metal blanks. The work was hard, tedious, and sweaty. To make their job easier, workers increased the downward force the press was capable of exerting by attaching heavy, spherical weights to the outer ends of the screw handles. These weights were nicknamed "fly balls," and workers either shoved them sideways or yanked ropes attached to them to set them in motion, adding their inertia to the downward force being applied to the metal blanks. Aided by angular momentum, these mechanized, spinning presses could strike a brand-new coin every couple of seconds.

Schwungräder come in sizes both small and very large indeed. The Railway Technical Research Institute in Japan recently developed one of the world's largest flywheels for electric power storage and generation; like the flywheels manufactured by Beacon Power, this two-meter, four-ton wheel banks electricity in the form of potential energy and converts it back to electricity on demand. Spinning at about 6,000 rpm, it's in essence a rechargeable battery that can be depleted and recharged forever, so long as there are people on earth who need it (Furukawa Electric, 2015).

At the other end of the scale is a type of flywheel that's truly flyweight: it's a single calcium ion whose intrinsic particle "spin" can be converted by a laser beam into vibrations or oscillations that store energy much like a potter's tournette. If you'd prefer a tiny flywheel that you can see at work, there's the oscillating steam engine built by Jerry Kieffer; the entire machine is less than a quarter inch long and would balance on a laboratory scale opposite two grains of rice. If you look closely when Kieffer's engine is in operation, you can watch two little flywheels spinning merrily away.

Whenever and wherever the historic forms of religious worship included a litany, humans soon found ways to simplify the recitations and/or repetitions of these crucial texts. Early Christians prayed with the help of knotted ropes to keep track of the number of times they spoke a particular text, and these crude, personal mnemonics are understood to be the forerunner of the rosary beads used in prayer by Catholic worshippers ever since the early Middle Ages. Early Tibetan worshippers, likewise, must have been just as prone to tune out mentally during the daily recitations of their devotions, and so they hit upon the idea to mechanize their worship by writing down the appropriate religious verses or mantras on scrolls and tucking them inside a container mounted on a hand-held spindle. The Tibetan "prayer wheel," thus, was not really a wheel but rather a spinning prayer cylinder. Its origin is obscure: most historians date mechanical, hand-held prayer wheels from the twelfth or thirteenth centuries, though primitive analogues may have existed earlier.

Wheel or cylinder notwithstanding, this clever device, says Tibetan scholar Dan Martin, was a "technological innovation ... occasioned by the exigencies of the spiritual life" (Martin 1987). The design hasn't changed much over the centuries. It consists of a round, hollow container about the size of a soup can, constructed usually of repoussé metal and mounted atop a wooden spindle on which the cylinder is free to rotate. A flick of the wrist starts the cylinder with its cargo of prayers turning, and a small, weighted ball and short string attached to the can's circumference shifts the moment of inertia outward and so adds extra rotational impetus to keep the "wheel" spinning.

While the prayers churn round and round, their words are slung outward into the cosmos, there to work their divine purpose just as surely if they had been chanted. It's a remarkably efficient form of devotional activity. Since each revolution of the wheel is believed to be the equivalent of a single recitation of the text it contains, and since thousands of different prayers can be loaded into a single cylinder, a dedicated supplicant (or possibly a dilatory one) is able to broadcast a full day's litany in little more than an hour. Hand-held prayer wheels are traditionally spun clockwise (i.e., "sunwise") an imitative movement that brings one's worship in empathic harmony with the sun's daily transit across the sky.

On a web site called soundcloud.com I listen to sounds recorded inside a Tibetan monastery. Deep, agonistic groans mix with the brighter noise of chimes and tinkling bells, the skittering clatter of small, hand-held prayer wheels. These are sounds meant to rouse the spirit. A project called "Sacred Spaces" has been documenting for preservation the sounds of spirituality, whether those are rhythmic chants, church organ music, or the dry, rattling chatter of hundreds of handheld prayer wheels. The recordings have been made as part of the larger "Cities and Memory" project, a massive collaborative attempt to create a contemporary global sound map—a snapshot, so to speak, of what the world sounds like in the first years of a new millennium. ("Remixing the world," as the projects' originators describe their work on citiesandmemory.com, "one sound at a time.")

I hear bells pealing from the Alexanderplatz in Berlin or the Basilica

San Salvatore dei Fieschi in Italy, raucous sounds of an Easter Sunday procession in the streets of Corfu, organ music from the church of St. Peter in the tiny British village of Allexton (population 58 and falling), calls to prayer from the minarets of the famed city of Istanbul. On the same site, I hear a chorus from Magdalen College, Oxford, celebrate the arrival of May, and I listen to the sound of a spinning prayer wheel recorded in Amideva Buddha Park in Kathmandu.

What does religion sound like? Like everything, like nothing. "How sweet it is," says the narrator in Umberto Eco's *The Name of the Rose*, "merely to sit in solitude and hold a conversation with God!" At the opposite end of the spectrum is the clattering, cheepy racket coming from the spinning cylinders of thousands of hand-held prayer wheels that fill the streets of Gyirong in the Tibetan Autonomous Region during the annual festival of Saga Dawa. Banks of slightly larger prayer wheels are tucked in alcoves in walls along the street where passers-by can flick them as they pursue their daily business. Other, much larger wheels—some the size of a walk-in freezer—are housed in gazebos and spun mechanically, chiming with each revolution of the cylinder to the background noise of busy city street. Prayer wheels have been installed as well on the roofs of houses where they are spun by the wind. Some are turned by flowing water in streams, still others spin with hot air currents rising from lamps or stoves. These self-turning wheels are believed to work just as well for those who spin them as for those who don't. Much as a virus spreads itself through a population, merely to breathe in air that has blown across a spinning prayer wheel is believed to cleanse one's sins and clear the pathway toward enlightenment, a kind of divine wind.

Wind is air in motion. It has mass, it has momentum, and like everything else with those qualities, wind can push other things around. You may be able to feel the air ruffle your hair or (if the breeze is strong enough) sway you about, but try as hard as you can, you just can't see it. Human skin is sensitive to three types of mechanical stress—pressure, shear, and torsion—and even the gentlest of breezes on, say, the back of one's neck is instantly detectable. Unseen forces like the wind are often believed to be next door to magic, hence the impulse to link the wind not just with magic but also with religion, eventually with prayer. It was the movement of the air, thought Sigmund Freud, that provided an image of spirituality, the spirit taking its name from the breath of the wind (Freud 1938).

In languages the world over, the words are the same: *Ruh* ("breath" and "spirit") in Hebrew and Arabic, *niya* to the Dakota and Sioux, *spirit* in English, from Latin *spiritus*, literally "breath" but more broadly "spirit" or, as the term came to be understood increasingly during the Christian era, "holy ghost" (Watson 1984, 327; Buck 1949, 1087). I remember a spinning set of chimes that appeared every December along with the rest of my parents' household Christmas decorations. Three brass silhouettes representing angels were hung beneath the blades of a small, horizontal fan; the flow of hot air rising from the candle flames spun both fan and angels round and round. As the angels twirled, thin rods dangling from their kneecaps struck a pair of bells mounted beneath, tinkling brightly. The contraption twinkled and tinkled away for as long as the candles kept burning, hour after hour. It was (from a child's point of view) a perpetual motion machine—charming, mysterious, and completely mesmerizing. Religiosity was in the mix as well, Christendom's iteration of a prayer wheel.

These Christmas-themed whirligigs were first marketed more than a century ago by the toy firm Adrian & Stock of Solingen, Germany. The Stock company didn't invent the mechanized icons, but they were the first to patent such toys and to upgrade them specifically as religious-themed "angel chimes," or *Engelgelaüte*. Stock's patents for a "Gelaüt" include just about every common feature of contemporary angel chimes, including flying angels with clappers dangling beneath and a lone, trumpeting angle perched on top . Fulsome descriptions of the device were published in 1906 in the magazine *Berliner Illustrierte*, and the toy was boxed lavishly as *Engel-Weinachts-Gelaüt, Posaunenchor mit der Geburt Christi*, or "Angel-Christmas Chime, Trumpet Choir with the Birth of Christ."

Much more elaborate Christmas-candle pyramids have been popular in Germany for centuries, where they come in all sizes ranging from tiny table-top playthings to the 50-foot tall *Weinachtspyramid* assembled each holiday season in Dresden. There's even a parody of angel chimes in the film "National Lampoon's Christmas Vacation." After drunken cousin Eddie accidentally smashes the fan vanes on the Griswold family's wooden *Engelgelaüt*, his daughter, the (formerly) cross-eyed Ruby Sue (played by Ellen Latzen), lights the candles hopefully anyway.

Civilized societies, sadly, are capable of great mischief, none more dispiriting than when channeling a new technology into novel ways to commit mayhem and murder. Electricity brightens houses and streets, but also spawned in the laboratory of Thomas Edison was the idea for a chair known as "Old Sparky." The electric chair was allegedly built as a more humane way to put criminals to death.[4] So, likewise, was wheel technology once converted to diabolical ends. During the early Christian era in Europe, for example, large, spoked wheels were turned into executioner's weapons where victims' bones were cruelly broken on a device called a Catherine[5] wheel—also known variously as "the breaking wheel," "the execution wheel," or—simply and more ominously—"The Wheel." It wasn't a one-off affair either. Torture and execution by wheels of one sort or another has been documented since antiquity. The sixth-century writer Gregory of Tours describes how criminals were sometimes crushed to death by driving over them with heavy-laden wagons (Spierenburg 1984, 71). Similar methods were practiced occasionally during the eighteenth century in North America as well as on the Indian subcontinent; a comedic version of execution by wheel even appears in the 1987 fantasy/comedy/adventure film *The Princess Bride*, when the hero Wesley, a farm-boy-turned-pirate, is tortured on a diabolical machine that uses wheel-driven pumps to suck out his life.

The most well-known type of executioner's wheel, the Catherine wheel, was not the actual means by which victims were killed but rather the name of the site where the killing took place. To read accounts of such executions chills the spirit. A person was first bound to a large, spoked wheel, then beaten with clubs or stones until his limbs were broken in multiple places. (One might ask: why use a wheel in the first place? Wouldn't chaining the victim to a wall or post be simpler? There seems no particular reason for fixing a person to a wheel to beat him to death except for the convenience of spinning the victim round from time to time for the executioner to get a new angle for administering his blows.)

Once the sufferer's bones had been shattered, his broken legs and arms were threaded in and out of the spokes, and more beating followed. At some point during this long, cruel, and unimaginably painful torture, death occurred. But even then, the grisly spectacle was not necessarily over. Often the shattered corpse was left on the wheel for days where it was picked over by scavenging birds. To be "broken on the wheel," as it was called, was perhaps the most brutal form of torture ever to come from human imagination; one victim, a Jew named ironically Bona Dies, endured the wheel for four agonizing days before dying. Even Jesus Christ on the cross was not so horribly tortured; the Savior had to suffer merely from nine till three before commending His spirit to God.

CHAPTER 6

Bullet Points

It is no good firing at 1,000 yards if there is not the remotest chance of hitting anything intended to be hit. Enter rifling.—W.S. CURTIS: "Long Range Shooting: An Historical Perspective"

Looking Down the Bore

Almost all James Bond adventure films open with the same visual sequence—an image of Britain's most famous spy seen as if by someone looking at him through the barrel of a gun. Bond himself is framed entirely within what appears to be the muzzle opening; he appears as a slim, dark-suited figure centered within a small circle of light. All around him looms the darkly luminous gun barrel; numerous sinister, curving lines spiral down its inner surface, like the current flow sweeping down a vortex. If a bullet had eyes, this is what it would see.

These twisting, shallow grooves inside the bore of a gun have a specific name. They're called "rifling" (from the middle/high German transitive verb, *riffeln*: "to scratch, ripple, or provide with grooves.")[1] The spirals may be visually arresting—similar patterns have sometimes been used to induce states of hypnosis—but they serve a deadly purpose. They squeeze the bullet tightly enough to set it spinning rapidly as it explodes down the barrel. Why would you want a bullet to spin? Because in spinning, it will acquire angular momentum, and once it acquires angular momentum it will be less prone to veer from its intended flight path.

But where did the idea come from that you could make lead balls fly with much greater accuracy if you somehow set them spinning? Regrettably, the name of the person (or persons) who first thought to cut fine twists inside the barrel of a smoothbore musket has been lost to history. A thorough search on the internet turns up little useful information and nothing specific. Some commentators give Leonardo da Vinci

119

Rifling of 9 cm. cannon used by Austria-Hungary forces during World War I (Petar Milošević).

the credit—or the blame—as the first person to conceive of rifled barrels. He's indeed a distinct possibility. Often, when Leonardo wasn't otherwise occupied in painting masterpieces or (possibly) sculpting, he was working on behalf of the bloodthirsty Borgias, sketching plans for a slew of then-undreamed-of weapons like giant crossbows and steam-powered cannons, not to mention armored tanks, machine guns, and submarines.

There are candidates besides Leonardo for the honor, however. According to one widespread theory, whoever thought of setting a lead slug to spinning did so to imitate the spiraling trajectories of bolts shot from medieval crossbows. Supposedly, those arrows were sometimes fletched at a slight angle to cause them to spin directly around their longitudinal axis while in flight, thereby achieving a greater measure of accuracy. "Gunmakers are inventive people," says David Westwood, the author of *Rifles: An Illustrated History of Their Impact*, "and even in the days of wheel lock they had noticed that archers had a trick or two up their sleeves worth thinking about. One of these tricks was to offset the fletching slightly so that in the air it would rotate about its longitudinal axis and thereby gain stability in flight" (Westwood 2005, 14).

Westwood's story makes sense. If warriors' arrows that spun in flight found their mark more often than those that flew without twisting, it would have taken no great leap of gun-makers' imaginations to suppose the same results might apply to speeding bullets. But is it true that spinning arrows are so much superior in accuracy? The history of offset fletching is just as difficult to pin down as the invention of rifling, so the influence of arrow-making on gunnery may be negligible if not nonexistent. Besides, the advantage of offset- over in-line fletching seems slight. One recent experiment (July 2011) that I consulted, for example, showed that offset/helical fletching resulted only in minor improvements in accuracy.[2] It's doubtful that offset fletching would have mattered enough to justify the extra time and greater precision required of medieval arrow makers to install it. It's harder still to believe that such minor benefits in accuracy would have sparked the imaginations of early gun-makers whose weapons were notoriously inaccurate for many other reasons.

There is, in the many different stories of the invention of rifling that I consulted, a bland sameness that suggests more than anything else that some fundamental ignorance is being papered over. In Wikipedia, for example, I read the confident assertion that "[b]arrel rifling was invented in Augsburg, Germany in 1498" (Wikipedia.org 2020). This same writer (writers?) notes additionally that "[i]n 1520 August Kotter, an armourer of Nuremberg, Germany improved upon this work" (Wikipedia.org 2018). That sounds fair enough. Unfortunately, however, the ultimate source of this information belies the confident tone with which it is written and published. The Wikipedia account derives from—and embellishes—a lone article by the firearms historian W.S. Curtis (of the Chapter 6 epigraph).

Curtis himself, in clear contrast to the later Wikipedia contributors who draw on his research, is reluctant to assign a date, place, and name to the first gun maker in history to cut a series of spiraling grooves inside the barrel of a long gun known forever henceforth as a "rifle." "Rifling is said," he writes (the passive voice conveys hesitancy) "to have originated in Augsburg around the year 1498 and certainly target practice with the arquebus was common by then. In the early 16th century, there are references to banning grooved barrels because they were unfair" (Curtis 2014, 5).

A separate online history (also sourced ultimately from Wikipedia) places the earliest rifle neither in Augsburg nor Nuremberg but many leagues to the south in Wien (Vienna), where it is said to have been the handiwork of an early sixteenth-century gun maker, Gaspard Kollner/ Koller. The information on Kollner/Koller comes in turn from a book first published in 1881 by W.W. Greener, *The Gun and Its Development*. Like Curtis, Greener is knowledgeable and judicious in his assessment of the known facts, but, like Curtis, he hesitates to assign the invention of rifling to a particular time, place, and person.

So, then, was it Kotter of Nuremberg, or Kollner/Koller of Vienna? The names all seem suspiciously echolalic. But whether it was Kotter or Koller or Kollner who first thought to "rifle" the inside of a gun barrel, numerous subsequent experiments proved beyond any doubt the superiority of the technique. The general principles and benefits of rifling were described fully in 1841 by the American gunsmith Henry Wilkinson, who specified that seven to fifteen spiraling grooves were to be cut into the bore of a gun barrel, dividing it into a sequence of "furrows" (the grooves) and "lands" (what remained of the formerly smooth surface). The purpose of the spirals, as Wilkinson documents, was to convert the random motions of a spherical ball into a uniform, stabilizing spin. As we'll see, however, an English schoolteacher named Benjamin Robins had exhaustively demonstrated the same principle more than a century earlier.

The Schoolmaster and the Military Engineer

So many of the men and women to whom we owe so much get so little credit in the history books. The eighteenth-century English mathematician Benjamin Robins is one such forgotten figure. Yet more than any other single man or woman, this one-time tutor of schoolboys is responsible—completely inadvertently, of course—for the lethality of modern arms. Between Robins' amateur fiddling with gunpowder and musket balls in the open fields near Woolwich (in southeast London) and modern sniper rifles capable of kills at distances greater than 2,500 yards, there runs a single line of descent. Yet unlike Alfred Nobel, the famous Swedish "merchant of death" who gave the world dynamite, few today know Robins' name. Worse, also unlike Nobel, while Benjamin Robins was steadily discovering ways to make guns ever more deadly at ever greater distances, he didn't get rich in the process. He died young, alone, half a world away from his homeland, fretting over a couple of trivial debts. (Military service in foreign lands was not an uncommon destiny for young men of that era who sought brighter futures and fortunes than they had been born to.)

Writing on the 25th of July 1751, from what he knew full well was his death bed, the brilliant math-teacher-turned-military-engineer spent his last few hours settling his financial accounts. To his embarrassment, he acknowledged a debt of £200 outstanding. Then, after disposing of a few other niggling bills, Robins entreated the officers of the East India Company to continue to support several men with whom he had worked most closely—John Brohier, John Call, and a servant named George Reynolds.

Over the course of his all-too-short life Robins brought scientific method to gunnery, a field that up to that point had been literally a hit-or-miss business. Mostly miss, in fact: European armies had been firing musket and cannon balls at each other for centuries, beginning well before the siege of Calais in 1350, but it took another four hundred years until Benjamin Robins set out to discover exactly what happened to those balls between the time they left the muzzle and hit—or more often missed—the target. It was Robins, for example, who first determined just how much influence the air itself had on a projectile in flight. Before he conducted and published the results of his extensive series of actual field tests with firearms, the effect of wind and air resistance (or *drag*, as it is now called) on something as small, dense, and speedy as a bullet was assumed to be negligible. As for the effects of drag when firing cannon balls, to worry about it seemed ludicrous: what possible influence could a passing summer breeze have on, say, a forty-pound chunk of iron? Such a great weight, and moving so fast! Even the great Isaac Newton had considered the effect of air on musket and cannon balls not worth worrying about.

Robins' relentless experiments and calculations showed just how wrong that thinking was. The arc of Benjamin Robins' life ran from a cradle near the banks of the river Avon to a gravesite by the curving coastline of the Bay of Bengal. He was born on the eve of Europe's Enlightenment in the old Roman town of Bath, the only son of a poor tailor. Both his father and mother were Quakers, and there is no record of Benjamin's ever having attended public school. When and where the lad picked up his extensive knowledge of mathematics, therefore, remains a mystery. But on the strength of his remarkable proficiency in that subject, wherever he had acquired it, the teenage Robins was able to leave his hometown of Bath and move to Cambridge where he soon made a small income by teaching students who were planning to enroll at the prestigious university but first needed help with their math skillset. (Robins seems to have conducted himself more like an SAT coach than a conventional classroom teacher, since his habitual manner of engagement with his students was one-on-one.)

While he was tutoring future Cambridge scholars, Robins began publishing papers regularly on mathematics and science. Right away this work attracted considerable attention, and he soon was recognized as a prodigy. To put this in perspective: at an age when young men and women of today are deciding whether to major in, say, political science or sociology, Robins was writing extensive papers for the *Philosophical Transactions of the Royal Society*, papers in which he detailed exactly where and how the top scientists and mathematicians of that era had got things wrong. In 1727, for example, the twenty-year old Robins published a correction to Isaac Newton's *Treatise of Quadratures*, while in another paper that appeared the following year, he took on another illustrious physicist/mathematician, Johan Bernoulli and convincingly refuted his impact theory of elastic collisions.

But the sedentary and largely cerebral pursuits of a scholar seemed insufficient to satisfy Robins' notion of a full life, and, according to one of his friends, the medical doctor James Wilson, Robins gradually abandoned teaching to follow career pathways that required (in Wilson's words) "more exercise" (Robins 1760). More exercise, indeed: after leaving university life behind him, Robins seems to have been a one-man construction crew, throwing his labors into erecting walls, building bridges, draining fens, taming wild rivers and rendering them navigable, carving out harbors, and constructing fortifications. Among Robins' many different engineering projects—and the one that is our chief concern here—was extensive fiddling with guns and the mechanics of gunnery.

How Robins developed an interest—in retrospect, it looks more like an obsession—in ballistics is not clear. A decade after he had begun teaching mathematics and publishing scholarly articles, Robins had applied for—and was for unknown reasons denied—a position on the faculty of the Royal Military Academy. The most likely explanation was that he was turned down because of his public advocacy for Tory politicians and politics. Whatever the reason that Robins did not receive a faculty appointment, that event seems to have been a personal watershed; the rebuff convinced him that any further progress in a teaching career was effectively blocked, and he turned his attention once more to engineering projects, especially—and this was something brand new—those of a martial nature.

Perhaps Robins' burgeoning interest in military affairs was intended to show up the faculty of the Royal Academy who had spurned him? In any case, the one-time math tutor and riverine engineer soon became a leading authority on military fortifications and ballistics, ultimately settling fully into a military engineering career. In 1750, he accepted a posting to Madras (now the city of Chennai), India, where he was appointed captain of artillery and engineer general for the British East India Company. Sadly, upon reaching his new home, Robins almost immediately contracted a "fever of unknown origin." (Was it malaria? Dengue?) In a matter of weeks, he was dead, at the age of 44.

He died, according to the testimony of a friend named Robert Orme, with his pen in his hand. A letter posted from a subordinate to the home office of Company summarizes Robins' last few hours. Even allowing for the more ostentatious eulogia of that era, one feels in the subordinate's words the force of Robins' character: "A noble and distinguished man, albeit his name is now almost forgotten in the towns whose defences he labored to make secure, and in whose service he spent his latest breath," Orme wrote. "No more shining example of single-hearted devotion to duty in the face of exhausting illness can be found in the whole range of Anglo-Indian history than that of Benjamin Robins" (Love 2015).

It is Robins' early investigations into the spin of projectiles, however, that chiefly concern us here. Ever since the use of gunpowder as a propellant became widespread in Europe, philosophers no less than military commanders wanted to know what happened to a cannon- or musket ball after it left the muzzle of the gun (Kelly 2004). Because the chief advantage of guns is that their use enables one to kill or maim an enemy combatant at some distance and so avoid personal injury to oneself, the two most important questions for a cannoneer involved distance and accuracy. It was proved relatively early on by experiments conducted by the Italian engineer Niccolò Tartaglia that a barrel elevation of 45 degrees gave projectiles the longest range. That much was easy; point your gun halfway up the sky and shoot. But obtaining accuracy proved a much more difficult challenge.

The trials young Benjamin Robins carried out with firearms were as diverse and ingenious as they were exhaustive. Reading through his extensive publications you get a picture of the man in full and his careful, judicious mind. When Benjamin Robins grabbed onto an idea it was like a pet Labrador mouthing a tennis ball; you have the sense neither would relinquish hold of the thing until it had been thoroughly chewed over. "I have made several hundred shots in very different seasons," he writes in his journal, "sometimes compared the trials made at noon in the hottest summer sun, with those made in the freshness of the morning ... and it was the same made in the night and in winter." But the one-time tutor of schoolboys wrote also with courtesy, especially when overturning Newton and what was taken at the time to be settled science: "From all that we have related, then, it appears that the theory of the resistance of the air, established in slow motion by Sir Isaac Newton, and confirmed by many experiments, is altogether erroneous, when applied to the swifter motions of musket or cannon shot" (Robins 1761).

Robins at first must have been astonished to see musket balls miss their intended targets by huge margins. His journals record that the balls typically strayed approximately one yard left or right for every seven yards they traveled forward—a rate of deflection that would have made aiming a gun at a particular target all but useless, dueling as low-risk an activity as carpentry. As soon as a ball left the muzzle, Robins observed, it become seriously unstable in flight, wobbling and tumbling end over end. Puzzling over the reasons for this inaccuracy, Robins guessed that a bullet was being hammered about violently and unpredictably as it sped down the barrel, and he guessed further that whatever lateral momentum it picked up would be continued on exit.

"The greatest part of military projectiles," he reasoned, "will at the time of their discharge acquire a whirling motion round their axis by rubbing against the inside of their respective pieces; and this whirling motion will cause them to strike the air very differently, from what they would do, had they no other but a progressive motion" (Robins 1761, 257). The trajectory of a spinning projectile, he concludes, "will constantly incline to that hand, towards which the revolving motion tends" (Robins 1761, 258). In what is surely a nod to Newton, Robins adds that "the same [swerving movement] is visible in tennis balls" (Robins 1761, 258). His thinking here was, in essence, entirely correct; his reasoning was so elegant, in fact, that it has been suggested that the modern laws of physics that describe the motions of spinning projectiles in transit through a fluid medium ought properly to be called the "Robins Effect," rather than the "Magnus Effect," as the principle has come to be known.[3]

Whether a bullet spun left or right, with topspin or backspin, Robins understood, was entirely up to chance; there would be no predicting its trajectory. But no sooner had he perceived the problem than he grasped the solution: if a lead ball could somehow be set spinning *regularly* along a central axis while it was in the gun barrel, it would acquire gyroscopic forces that would help stabilize it. The projectile would spin like a top, maintaining the same orientation throughout flight. "This led," says small arms historian Brett Steele, "to his trials with rifled barrels, the results of which were published in a paper read before The Royal Society in which he forecast that the first State to adopt the rifle generally will acquire a superiority unequalled since the invention of gunpowder" (Steele 1994, 365).

"Of the Nature and Advantage of Rifled Barrel Pieces" was read before the Royal Society on July 2, 1747; in it, Robins demonstrated the foolishness of previous theories of why projectiles launched from rifled firearms flew straighter than those shot from smoothbore weapons. (One widely respected theory at the time—which Robins thoroughly but courteously discredited—was that rifling made a bullet "bore" its way straight through the air much as a screw bores through a pine plank.) The real advantage of rifling, Robins correctly insisted, was that it kept the leading part of a projectile always facing nose to the front, stably rotating around an axis parallel with its initial trajectory.

"It should follow," he proposed, "that the same hemisphere of the bullet, which lies foremost in the piece, must continue foremost during the whole course of its flight." To test his theory Robins once again hit on a sublimely simple method. He replaced lead bullets with facsimiles made of "soft, springy wood" which could be marked and then inspected after impact: "[F]iring the piece thus loaded against a wall at such a distance," he observed, "as the bullet might not be shivered by the blow; I always found, that the same surface, which lay foremost in the piece, continued foremost without any sensible deflection, during the time of its flight."

Robins closed his dissection of rifling with a vision of the future of warfare as clear as stars seen through the renovated lens on the Hubble telescope: "whatever state shall thoroughly comprehend the nature and advantage of rifled barrel pieces [and] ... shall introduce into their armies their general use ... will by this means acquire a superiority, which will almost equal any thing, that has been done at any time by the particular excellence of any one kind of arms" (Steele 365). He was right; in fact, it was owing only to the difficulty of manufacturing rifles in mass quantities that kept smoothbore firearms on European battlefields until the end of the nineteenth century.

A Poetics of Ballistics

I'm standing in a tiny wine shop on the corner of an alley in Moscow, Idaho. I'm the only customer, and soon I strike up a conversation with the young man behind the counter. He is reading *Medieval People* by Eileen Power, first published in 1924. I tell him I know that book, and soon we are deep in conversation about Marco Polo and Madame Eglentyne, about Dante, the crossbow, and the stirrup. He says he is a student at the University of Idaho, working toward a bachelor's degree in history, paying for his education with the G.I. Bill. I tell him I also went to the same school on the same bill, a lifetime ago, after two years in the Marines.

We talk about military life. He's a former sergeant in the Army, six months back from a second deployment to Iraq. Does he know Paul Fussell's books on World Wars I and II, *Wartime* and *The Great War and Modern Memory*? He does. We gossip about the afterlife of a soldier. I tell him about a conversation I once had with a member of the faculty at Emory University in Atlanta where I used to teach. Professor of English Lucas Carpenter was a former infantry officer in Vietnam. One morning as we walk across the campus, Carpenter and I pause to take in the beauty of a southern spring. All around us, dogwoods and azaleas are exploding with color—red, white, flagrant shades of pink. The sun is brilliant, the air soft and sweet. The landscape hardly seems real; this, I think, could pass for Eden. So I am surprised to hear both regret and bewilderment in Carpenter's voice when he tells me he cannot look at this space without at the same time wondering where he would place a machine gun if some day, God forbid, it were necessary to defend it. "With me it's garbage cans," says the young man behind the counter in the wine shop. "I can't see a pile of trash without flashing *is it a bomb?*"

When I began to research spinning bullets, I had assumed that belletristic writings on a subject so gruesome would be few and far between. Not so, I discovered. On the website *poemhunter.com*, for example, I found dozens of poems about bullets; a recurring technique among contemporary bullet-poem-poets is to build variations on the common sexual trope, Cupid's arrow of desire. In this connection I remembered the lyrics to Bon Jovi's well-known rock song of 1986: "Shot through the heart/And you're to blame/You Give Love a Bad Name."

Other bullet poems took up historical subjects. "A Bullet for Lenin," by Patti Masterman, commemorated an attempted assassination of Russia's revolutionary leader in August 1918. Still others were experimental; one nineteenth-century poet named Francis Bret Harte used prosopopoeia to celebrate the thoughts and feelings of a speeding bullet: "O joy of creation/To be!/O rapture to fly/And be free!" And a few were risible and sad simultaneously: Hasmukh Amathalal, for example, composed verses called "Bullet Proof" because he was angry that several police officers in Mumbai had died because their inferior body armor didn't protect them against terrorists' bullets:

> You have laid your life in vain,
> Worry was not for self main,
> But the way it was to be put off.
> Only because of jacket bullet proof.

I turned next to Herman Melville and Robert Frost. In the short lyric poem "Shiloh: A Requiem (April, 1862)," for example, Melville reflects on a visit to the site of a Civil War battle whose horrific carnage stunned both North and South. He first describes the now-quiet landscape where swallows, "skimming lightly ... over the field in clouded days," then imagines all the dying soldiers who in their last moments of consciousness reflect on the lies they have killed and died for:

> Foemen at morn, but friends at eve—
> Fame or country least their care:
> (What like a bullet can undeceive!)

Robert Frost takes a more dispassionate attitude toward a bullet in the early poem "Range Finding," published in 1916 during World War I. This strange poem—ironically, given its morbid subject matter—was composed as a sonnet. It looks at the trajectory of a lone bullet as it might be experienced from extra-human points of view.

> The battle rent a cobweb diamond-strung
> And cut a flower beside a groundbird's nest.

Even if this bullet never found "a single human breast," it does a great deal of collateral damage, so to speak. It slams through a spider's web and partly slices the stem of a flower, simultaneously "dispossessing" a butterfly that happened to be sitting on the blossom sucking nectar and confusing a spider that interpreted the sudden trembling of its web as the struggles of a newly ensnared fly. When compared to its destruction of the lives of flowers and insects, that the bullet might next kill a human seems incidental.

More recently, Brian Turner, a former Army sergeant, published a book of war poems entitled *Here Bullet*. The title poem of the collection takes the point of view of a soldier in combat, brazenly summoning a bullet to strike him: "If a body is what you want/then here is bone and gristle and flesh.... I dare you to finish/what you've started" (Turner 2005). Turner is a veteran of the Iraq war, where he served as a sergeant in an infantry unit; his picture of small arms warfare is immediate, blunt, and accurate. "Unaccountably beautiful" (which is how Yale's Michael Warner has described Melville's "Shiloh") are not words that came to mind when I read Turner's war poems (Warner 2020).

Even Frost's much bleaker stance still aestheticizes its subject; compare "a sudden passing bullet shook it [a spider's web] dry" with "triggering my tongue's explosives for the rifling I have/inside of me, each twist of the round spun deeper...." Turner's account is gruesome but likely accurate. On an online technical forum on bullet spin, I found the following entry by a writer "SW" who commented on what some of the participants were calling the "buzz saw" or "router" effect when spinning bullets slammed into flesh at hypersonic velocity: "Do you remember the 'shot in the stomach, but the bullet wound up in the guy's shoulder' stories that came back from Vietnam? Guess how the little 68 fmj [68 gram weight, full metal jacket] bullet got there? It spun its way up there. With retained rpm" (AccurateShooter.com 2008).

One of the first mass-manufactured bullets to make use of spin was the "Minié Ball," so designated after one of its developers, a nineteenth-century French army captain named Claude-Étienne Minié. If you Google the ball that made Minié a household name on battlefields of that era, you'll see that this so-called "ball" didn't look like one, not even remotely. It was a cylindrical slug. But the bullet was nevertheless called a "ball" out of inertia on the part of its manufacturers, because for centuries beforehand, gunners used round balls cast from molten lead. (Modern military nomenclature still labels small arms bullets "ball ammunition.")

Early rifled pistols and muskets that fired spherical lead balls were notoriously clumsy and inaccurate; their problems were twofold, as Benjamin Robins had proved beyond doubt. If the fit of ball to barrel was loose, the projectile would tend to ricochet off the sides of the barrel when the gun was fired, seriously compromising accuracy. A tighter fit against the grooves cut inside the barrel made the ball's trajectory more predictable. But a friction fit made it difficult to shove the ball quickly down the barrel, then snug it up against the charge of powder. What good was accuracy if you couldn't reload your gun fast enough on the battlefield to get off multiple shots?

A Minié ball, in contrast, was a tiny bit smaller in diameter than the rifle bore, and so it slid easily down the barrel. It had a round nose and a skirted, hollowed-out back end with a small iron plug fitted inside it. (This plug was soon discovered to be useless and was later omitted.) When the powder exploded, the sudden pressure behind and inside the skirt blew it outward so the rearmost part of the bullet now fit tight against the spiral grooves inside the barrel. This increased pressure and greater contact area with the rifling substantially increased the bullet's spin rate. The new rounds were more consistent; one shot was just like every other. More spin and more consistency translated immediately into increases in both range and accuracy of a rifle; indeed, the effective range of a Minié type bullet was as much as 500 yards, about twice the effective range of ball ammunition. ("At longer ranges," says Justin Stanage, "beyond about 100 yards, the lone man with a smoothbore musket is not likely to hit a lone enemy target" [Stanage 2000]). The lethality of the Minié bullet was such an improvement—though "improvement" sounds like entirely the wrong word here—over earlier types of ammunition that it has been said to have accounted for as much as 90 percent of the battlefield casualties during the American Civil War.

Puff the Magic Dragon

Richard Jordan Gatling took a long time to find his niche. As a young man living in antebellum North Carolina, Gatling dreamed up a novel, screw propeller drive to power riverboats up and down southern waterways. His invention substantially outperformed older paddle drives, but when he arrived at the patent offices in Washington, D.C., to stake claim to his invention, he discovered that a Swedish locomotive designer with a similar idea had beat him there by a matter of days (Jordan 1986).

For most of the rest of his twenties Gatling ran through the gamut of occupations then open to young men without much formal education, trying out many but settling on none; at different times he was a dry goods clerk, a law student, a schoolmaster, a farmer, and manufacturer of agricultural implements. When he wasn't behind a counter or in front of a classroom, Gatling was producing and selling a variety of agricultural machines he had built for planting, thinning, and chopping crops on his family's plantation on Carolina's coastal plains. On a riverboat business trip in the winter of 1845, however, the rich, young entrepreneur became infected with smallpox, fell gravely ill, and nearly died. His unsolicited dance with death spurred Gatling to pursue a degree in medicine. For once he stayed the course, and three years later he received his M.D. from the Ohio Medical College.

The new career soon brought young Doctor Gatling suddenly into contact with the appalling realities of the Civil War. Seeing that many wartime casualties came not from bullets but from a toxic brew of sicknesses contracted while tens of thousands of men were encamped in unsanitary conditions in large numbers, Gatling reasoned thusly: a soldier with a rifle kills one man at a time. Killing a lot of men, therefore, takes a lot of soldiers. But what if a single gun somehow could fire many bullets rapidly? Battles would then require fewer men to fight them; ergo, fewer men overall would die. The solution was obvious if somewhat perverse: to make war more humane, make guns more lethal!

Bizarre as this logic may sound now, Gatling's thinking was considered reasonable at the time. (And really: it differs little from the Cold War Era policy of "mutually-assured-destruction" adopted by the United States and the then Soviet Union.) A contemporary British newspaper, for example, printed this surprisingly enthusiastic endorsement of Gatling's new gun: "The general use of the formidable weapon will tend to diminish the barbarity and actual carnage of warfare, as its known relentless certainty of execution will help to prevent wars and thereby aid in keeping the peace of Christendom" (Jordan 1986). Would it had been so, alas!

Gatling's first design for a gun to "civilize" warfare employed a spinning mechanism that fed cartridges one at a time into a single gun barrel. In a bizarre twist on Isaiah's dream that the people would one day beat swords into plowshares and spears into pruning hooks, Gatling fashioned his gun more-or-less straight from a planting machine he had once used on his farm to sow rice seeds (Parramore 2006). This first-generation rapid-fire weapon looked like a hybrid revolver-rifle. But the invention that was ultimately to bear Gatling's name looked much different. It was, in essence, one gun made up of many separate guns, a kind of e pluribus unum of firearms. The most difficult technical problem with his (and other) earlier, rapid-fire guns was metal distortion caused by heat buildup. The more rapidly a revolver-rifle was fired, the more heat accumulated in the barrel, swelling it. Eventually the weapon got so hot bullets jammed, and it ceased to function.

Gatling's solution to the problem of thermal overload was as ingenious as it was diabolical: take the heat from multiple bullets fired down a single, fixed barrel and distribute it evenly among *multiple* barrels, all spinning round the same axis. He gathered six—later ten—separate gun barrels and mounted them in a ring around a central shaft. Barrels and shaft, in turn, were connected as a unit to a hand crank; bullets were loaded, hammers cocked and triggered automatically as the crank was turned and the whole apparatus spun round. The clerk-turned-doctor-turned-arms-designer had the world's first rapid-fire weapon ready for testing in 1862. His device was downright elegant. Multiple barrels never overheated; the gun never jammed. Rate-of-fire comparisons with conventional weapons of the era were instructive—and alarming. The standard Springfield Model 1861 rifle in use by many Civil War soldiers could fire from two to four rounds a minute, depending on the dexterity of whoever was using it; in the same amount of time, even the earliest of Gatling's multiple-barreled weapons could send downrange *hundreds* of bullets. Advancing troops faced a wall of deadly fire; there was nowhere to hide.

During the next several years Gatling kept tinkering with his gun's design, and by the time the War Between the States was over, he had considerably improved—if that is the right word—on his original creation. One year after the end of the war, the Colt Fire Arms Company bought patent rights from Gatling for a multiple-barreled gun that fired 1,200 rounds per minute. A later design more than doubled even that torrential rate of fire; it sent a bullet flying through every two-foot square space. Justifiably impressed, the U.S. War Department bought a hundred of the guns; even larger orders from England, France, Germany, Italy, Russia, and Turkey soon poured in (Stephenson 1979). George Armstrong Custer himself was offered the use of three of Gatling's guns, but the colonel chose not to take them with him as he rode out with his battalion on the morning of June 25, 1876.[4]

In the middle of the twentieth century, General Electric built an electrified version of Dr. Gatling's gun; during the Vietnam War, some were mounted in militarized DC-3 aircraft for use as airborne gunships. Fitted with three electrified Gatling-style "miniguns," these aircraft (nicknamed "Puff the Magic Dragon" after the popular folk song of the time) were able during one three-second burst to fire one bullet into every square yard covering half the length of football field. "Suppressive fire" indeed. Similar weapons are currently fitted to most naval warships, also to military aircraft like the A-10 "Warthog," a ground-support plane specifically designed for the sole purpose of carrying a Gatling-type gun, the GAU-8 "Avenger" rotary cannon. Gatling's spinning creation thrives also in the movies, where its ferocious firepower allows for spectacular visual effects. In the 1987 movie *Predator*, for example, Jesse Ventura, the pro-wrestler and (later) governor of Minnesota, plays a soldier-of-fortune named Blain Cooper who shreds half a forest when he fires his rotary, GE M134 Minigun at the predatory alien who is stalking him.

Faster than a Speeding Bullet

The physical forces acting on a spinning, high-velocity bullet are no different from the ones that act on ice skaters or curve balls, though they have somewhat different effects. Let's look briefly at them. The first is something called gyroscopic drift, which is basically what made your toy gyroscope wander slowly across the table after you started it spinning. Both toy and bullet tend to move in the direction of rotation, though not very much overall. Data from the *NRA Firearms Sourcebook* cite a gyroscopic swerve of about one-and-a half inches left or right (depending on the orientation of the twists in the barrel) over 500 yards. That's an effect that can be easily compensated for when sighting in a firearm. It's the same with the Magnus force, usually insignificant. The directional stability gained by spinning the projectile (as Benjamin Robins proved) is far more important.

Next comes into play a different kind of sideways movement, a result of something called the Coriolis Force.[5] A bullet or an artillery shell travel naturally in a straight line. But because any gun is located on the surface of a spinning planet (unless you're standing on the North or South poles when you pull the trigger), the trajectory of the bullet will also have a *sideways* vector component. Deer hunters and Olympic competitors don't have to worry about their bullets being flung sideways, however; at 40 degrees latitude, the angular velocity of the rotating Earth will curve a bullet less than a tenth of an inch for every 125 yards of forward travel. Even at ranges as great as a thousand yards, says Hornady ammunition manufacturer spokesman Steve Johnson, variables such as wind, air density, and "good old-fashioned shooter variance will overshadow any Coriolis force that may be at work" (Sagi 2017).

That's not the case, however, when it comes to big, heavy, long-range guns such as cannons, howitzers, and rockets that are mounted on platforms and aimed at targets situated many miles distant. In such cases, the Coriolis Force matters a great deal. "Artillery," continues Johnson, "is a different game [from small arms' ballistics], trajectories are extreme, and time of flight can be measured in minutes, providing much more opportunity for the Earth to rotate under the projectile in flight." Toward the end of World War I, for example, German gunners who aimed their cannons at Paris from 75 miles away had to allow for a drift off-target of about half a mile (Collins 2009).

Several other things affect the trajectory of a bullet. Among them are drag, gravity, and angular velocity. Of these, the effects of gravity are simplest. I'm sure like me you were taught in school that a bullet dropped vertically would strike the ground at the same time as one that was fired simultaneously in a horizontal direction. The idea is that gravity would cause both bullets to be accelerated downward equally independent of any horizontal motion they had (or didn't have). This piece of scientific wisdom has been around for so long that it was probably inevitable that one day somebody would put it to the test. On October 14, 2009, the Discovery channel's *Mythbusters* team, Adam Savage and Jamie Hyneman, did the physics and ran the experiment. The two bullets struck the ground within 40 milliseconds of each other, close enough for Jamie and Adam to declare the myth (and so Isaac Newton) confirmed. More recently, physicist and television personality Neil de Grasse Tyson tweets the following: "A bullet fired level from a gun will hit ground at the same time as a bullet dropped from the same height. Do the physics" (Tyson 2010).

Drag and angular velocity act on a spinning bullet just as on a baseball. The former slows it down, the latter confers directional stability.

But saying that a bullet spins extremely fast doesn't do justice to the physics. How fast a bullet spins depends on the twist rate (or number of grooves cut into the barrel) and on the bullet's translational (linear) velocity. The two are interdependent—a bullet traveling down a rifle barrel at 3000 feet per second will have twice the spin rate of one moving only half as fast.

The more spin the better, at least in theory; the faster a bullet spins, the greater its angular momentum and the less likely it will be bumped off course by cobwebs and flower stems. But there's an upper limit: it's possible for a bullet to spin so fast that the combination of rotational acceleration and heat generated from friction will cause it to disintegrate in flight. In this case the bullet leaves the muzzle but never strikes home. It simply vanishes into thin air, like the squadron of naval torpedo bombers that disappeared mysteriously while flying over the Bermuda Triangle.

People who gather on online forums to discuss high-end ballistics seem to be wary of spin rates above 250,000 rpm. But much depends on any given bullet's composition and jacket design. Digging into the subject I came across this shooter's tale of bullets that blew apart in midflight: "No shade, 85 degrees, 25 shots into a 45-shot ring in an F-class practice match. No impacts into the berm and after too many of these unexplained disappearing bullets ... went over to the short range [and] shot a couple into a target at 25 yards and witnessed jacket separation and molten lead spray on the target" (forum.snipershide.com 2010).

Drag, the final force acting on a speeding, spinning bullet is crucially important, and to understand it we'll have to return one last time to the experiments of Benjamin Robins. Robins' most widely known invention was the ballistic pendulum. This was, as its name suggests, a simple pendulum he positioned in front of the muzzle of a gun. When the gun was fired, Robins measured the distance the hanging weight was swung backward after being struck by the speeding ball. This ingenious mechanism not only provided a direct measure of the bullet's inertia, but also, since the weight of the bullet was a known quantity, Robins' device made it simple to calculate (by Newton's second law of motion, $F = ma$) the speed at which it left the muzzle. The concept of the ballistic pendulum was simplicity itself—so simple, in fact, that one wonders why it took gunsmiths so long to make use of it.

But Robins wasn't content with simply determining how fast bullets were traveling when they left the muzzle. He realized that if the same pendulum were placed at various distances downrange from the gun, its different deflections could be used to calculate a projectile's loss of velocity over the course of its flight. Robins' logic, as always, was pristine: once a bullet exits a gun barrel, the force of gravity pulls it downward in a parabolic curve. But the only force that could impede the projectile's forward progress was whatever resistance it met from the air through which it passed.

The results of Robins' experiments set conventional thinking about air resistance on its backside. By taking comparative measurements with his pendulum over different distances, Robins proved that Newton had significantly—no, grossly—underestimated the resistance (or *drag*) that air causes on a rapidly moving projectile. His ballistic pendulums showed that Newton not only was wrong, but wrong by a lot. Writing with characteristic modesty (and with deference to Sir Isaac, then president of the Royal Society), Robins concluded that his tests demonstrated "an enormous resistance of air to swift motions much beyond what any former theories have assigned" (Robins 1761).

What could be a better reward for a young scientist than to correct the great Isaac Newton? How about being shrewd enough to see two hundred years into the future? While he was documenting the results of countless bullets banging countless times against pendulums, Robins detected a strange, entirely unexpected trick of nature. His early experiments showed clearly that the air resistance encountered by a projectile increased steadily with increasing velocity. This was as one might expect. Drag, his calculations showed, was twice as hard to overcome when moving at 400 feet per second than when moving half that fast. Much to Robins' surprise, however, the linear correspondence between speed and drag went only so far. Once bullets moved at velocities of about 1,100 feet per second, his data showed, there appeared a sudden, dramatic increase in air resistance. This additional drag was as unexpected as it was puzzling; it was as if bullets encountered an invisible threshold whenever they were accelerated to a velocity—it's best to use Robins' own words here—"nearly the same with which sound is propagated through the air" (Robins 1761).

Robins published the results of his final experiments with projectiles in 1746. It would not be until another two centuries had passed that Air Force Captain Charles Yeager flew a rocket-powered aircraft safely through what young Benjamin Robins had so very long ago perceived and documented as the "sound barrier."[6]

Interchapter II

Making Iron Come

At some time in the second century BCE, the merchant Lu Buwei of Qin commissioned three thousand scribes and scholars to write down all that they had learned. When they finished, *Master Lü's Spring and Autumn Annals* compressed all knowledge into some 100,000 words; those who read the book could acquire the store of human wisdom in less time than it takes to read Mark Twain's *Adventures of Huckleberry Finn*. There are words on the necessity of respect for civil arts, words for trade, for conservation, for filial piety. Among the compendium of things worth knowing, one scribe pens this: the lodestone[1] makes iron come or it attracts it. There it was again: things moving, set in motion without being touched. What was going on?

If you wanted to name *the* most important scientific breakthrough of the nineteenth century, the discovery of the kinship between electricity and magnetism by the Danish natural philosopher Hans Christian Oersted would be as good a choice as any. Born almost thirty years before his more famous name-twin, the beloved writer of children's fairy tales Hans Christian Anderson, the scientific Hans Christian was a Danish polymath whose research interests ran the gamut from the kitchen to the chemistry lab. (In 1819, he isolated the alkaloid compound piperine, which gives black pepper its flavor and bite, and in 1825, he figured out a way to produce a brand-new metal, subsequently called aluminum, from the common medicinal astringent known popularly as alum.)

On April 21, 1820, Oersted was lecturing on electricity and magnetism to a group of students at the University of Copenhagen when he decided to add some drama to his presentation. Fiddling on stage with a battery, some wire, and a magnetic compass, Oersted noticed that his compass needle could be repeatedly deflected from its usual northerly orientation by switching an electric current from a nearby battery on and off. It's not clear whether Oersted's audience was aware that they had suddenly become witnesses to one of the greatest discoveries in the history of science; undergraduate students in large lecture halls being undergraduate students in large halls, probably not. But to Professor Oersted, the conclusion was obvious and inescapable: a live electric current somehow was producing a magnetic effect! (The professor's spontaneous showmanship may have surprised him even more than it surprised his students: according to one possibly apocryphal story, Oersted had constructed his battery-and-compass apparatus to demonstrate that electricity and magnetism were *independent* phenomena.)

Oersted's experiment pointed to a deep unity in nature that had heretofore merely been theorized by wishful philosophers from Plato to Immanuel Kant. (A basic unity which is still being pursued, in our time, in the form of the quest for the so-called Theory of Everything, or TOE.) After conducting additional and more elaborate experiments with electric currents, Oersted showed conclusively that *any* flow of electricity through *any* wire invariably created a magnetized space around it; this "magnetic field" (he didn't call it that) spiraled round and round the wire much like a snake constricting its prey. For discovering the singular, elemental force we now call electromagnetism, Oersted won a prize from London's Royal Society and a cash sum of 3,000 francs (about $75,000 in current dollars) from the French Academy.

Once it had been proved that the flow of an electric current could induce actual magnetic effects, it was not long before scientists and inventors began to experiment with electrically induced magnetic fields that could set things in motion without ever touching them. The British mathematician Peter Barlow, for example, in 1822 mounted a wheel whose rim was fitted with star-shaped metal points that dipped in and out of a pool of electrically charged mercury situated between the poles of a horseshoe magnet. As the wheel turned, the points on the wheel alternatively completed and then broke a battery-powered electrical circuit, and the intermittent surges of current effectively generated and then collapsed a magnetic field inside the stationary field between the poles of the horseshoe. The interaction of these two different magnetic fields in effect created pulses of energy sufficient to cause the six-inch wheel to rotate continually. In theory at least, the little wheel could have run indefinitely.

Almost two centuries later, Barlow's wheels can still be seen silently turning round and round in high school physics labs or You-Tube videos. (You can find them for sale online, save that modern versions, unlike the original, do not make use of the highly toxic substance, mercury.) Barlow deemed his apparatus to be of negligible practical use;

even the slightest resistance would stop the thing from turning, and he soon turned his attention to other projects that seemed of more obvious benefit to humanity.

Instead of fiddling with motors, Barlow made his bid for an enduring legacy by laboriously compiling tables of the factors, squares and square roots, cubes and cube roots, and hyperbolic logarithms of all numbers from one to 10,000. At the time, or so he must have thought, to have all these numbers and relationships ready at the fingertips would be an invaluable and lasting contribution to the world. In retrospect, the choice seems a poor one. In an age of electronic calculators and computers, math reference books like *Barlow's Tables* have become functionally useless. A few copies of his magnum opus can still be found in libraries, musty and unsought.

Attitudes toward electromagnetism were different in the laboratory of the British scientist Michael Faraday. By the time Faraday settled down to work, it was widely known that an electric current flowing through a wire produced magnetic effects, just as Oersted and Barlow had shown. It was widely assumed, therefore, that both lodestones *and* current-carrying wires produced forces that pushed or pulled along straight lines. But Faraday imagined a different picture: he saw those magnetic forces in terms of invisible curves, circles, and spirals. To distinguish these mysterious powers from the common push-and-pull dynamics of everyday life, Faraday chose for them a new name: *field* (from Anglo-Saxon *feld*, a word carrying the notion of something "spread out"). Further experiments showed something else crucially important. *If* a wire were turned back on itself to form a loop, and *if* that loop were then placed inside a second, stationary magnetic "field" at the same time an electric current was passed through the wire, a force appeared that tended to rotate the loop around its central axis.

That was the key: spin. It had been known for millennia that anything that spins produced a useful form of energy called torque, which is simply the twisting force exerted on anything that rotates about an axis. It's the torque, for example, that your wrist exerts on the doorknob that provides the force to withdraw the latch, and it's the torque supplied by your foot on the pedals of your bicycle that propels you and the machine down the road. For centuries, torque had worked its magic in everything from bow drills to water wheels. Now, Oersted had raised the curtain on more than a lecture hall curiosity.

Two centuries later, the by-products of his experiment can be seen all over the place, from giant industrial lathes to the hand-held drills your dentist uses to ream out a cavity. The torque produced by interacting electromagnetic fields is necessary even to gasoline-powered machines such as the 12-cylinder Ferrari engine that has been described lovingly as "the most soulful mechanical invention since the dawn of the Industrial Revolution" (Harper 2017). Yet the Ferrari would be dead along the roadway were the fuel in those twelve soulful cylinders not ignited thousands of times a minute by electrical energy created by a spinning alternator.

Oersted may have been first to discover the correspondence between electricity and magnetism, and Faraday may have created the first electromagnetic rotation apparatus. But much of the credit for all subsequent and fantastically useful electric devices—from the AAA battery-powered toothbrush on your bathroom countertop to the Union Pacific Railroad's DDA40X, a 260-ton electro-motive freight hauler known to railroad workers as "Big Jack"—goes to a much less celebrated 19th-century Slavic mathematician named Moritz Hermann von Jacobi.

Born in 1801 in Potsdam (near Berlin), von Jacobi studied architecture at universities in Berlin and Göttingen, then went to work for Prussia's King Friedrich as a government designer of public buildings. During his thirties he leapfrogged from Göttingen first to Königsberg (now Kaliningrad, Russia) and then in 1835 to Dorpat (now Tartu, Estonia) where he received an appointment as a professor of civil engineering. But von Jacobi's real interest lay neither in architecture nor even engineering but in the pure sciences. No sooner had von Jacobi settled into his new digs at Göttingen than he abandoned engineering for the laboratory. The young architect was especially intrigued—obsessed might be a better word—by the then new natural phenomenon of electromagnetism.

The mysterious symmetry between electric and magnetic forces occupied von Jacobi for the rest of his life. He seems as well to have been afflicted with a case of wanderlust, for as von Jacobi aged, he carried out a personal *Drang nach Osten*, the scientist and his laboratories drifting ever more eastward until he eventually set up housekeeping in St. Petersburg and obtained full Russian citizenship. He died at age 72 near the domes of the old Trinity Cathedral in that city, a couple thousand kilometers away from his fatherland.

Von Jacobi was among the very first persons to grasp that a wheel that spun by electrically-generated magnetic force could be made into a source of dependable power—that is to say, a proper motor. Oersted's wire and Barlow's wheel had been mere amusements, bagatelles; Faraday's spinning loops were not much more practical. They were like those executive ball clickers (sometimes called Newton's cradles) that physics teachers show to students to illustrate laws of inertia. From the very beginning, however, von Jacobi seems to have envisioned an electrical device that would be powerful enough to be do some serious moving and shaking. It's only a slight exaggeration to say that his lifelong obsession with electromagnetism made possible the modern world.

Gifted with a mind that was both practical and theoretical (one biographer called him the Leonhard Euler of the nineteenth century), Moritz Hermann von Jacobi spent more than three decades making an assortment of mind-bogglingly different and successful gadgets. His creations ran the gamut from printing plates to motorboats and even to exploding undersea mines. But whether he was building boats or inventing ways to blow them up, the common denominator among his inventions was that all of them worked by means of electricity. In 1839 in St. Petersburg, for example, von Jacobi installed an electric telegraph line that ran from the Tsar's Winter Palace to the general staff headquarters across the square; from 1842 to 1845, the line was extended as an underground cable that ran about 15 miles from St. Petersburg to Tsarskoe Selo. (To put von Jacobi's work on the telegraph in historic perspective: it was not until 1844 that Samuel Morse electrically transmitted *What hath God wrought* to his assistant, Albert Vail, 44 miles away in Baltimore.)

At about the same time that he was demonstrating the potential usefulness of galvanic electroplating and telegraphic communication, von Jacobi turned his attention to warfare. Now working for the Tsar as the director of what was then a kind of Russian equivalent to the modern House Armed Services Committee, he was tasked with improving naval defenses in the Crimean War effort. As one might expect, von Jacobi's solution once again involved the flow of electrons. In 1853, he stuffed thirty pounds of explosives into a metal container and anchored it to the bottom of the Baltic Sea; an electric cable connected a mercury switch sensitive to contact on the mine to a galvanic cell (or battery) installed on the distant shore. When the switch was bumped, electrons flowed, and the bomb blew up.

Von Jacobi's undersea "torpedo" (as mines were then called) proved greatly superior to contemporary American weapons that had to be set off by men scanning the sea from observation posts on the shoreline. Those primitive devices worked about as well as the traps Wile E Coyote sets out for the Road Runner. Von Jacobi's diabolical brainchild, on the other hand, detonated instantly on contact with passing vessels, and, unlike contemporary floating bombs that had to be triggered by onshore observers, these undersea weapons functioned automatically and lethally both at night and in fog as well as in the daytime. They were the world's first legitimate smart bombs, greatly superior to those being sold to the Russians at the same time by the Swedish industrialist, Immanuel Nobel.

Nobel's bombs typically posed greater hazards to the men who laid them than to the ships they were designed to sink; his production line lacked sufficient quality control, and the bombs that rolled off it often exploded while they were being set—if, that is, they exploded at all. (Immanuel's son Alfred, the chemist known famously first for inventing dynamite and, later, for endowing the prestigious and much-coveted international prizes that bear the family name, proved to be somewhat more adept than his father when it came to blowing things up.)

Von Jacobi was not the very first person to construct a functional electromagnetic motor. But he was the first to build one that could be put to any real use. Once it was proved that significant magnetic forces could be created and dispelled instantly just by switching an electric current on and off, it was "not difficult" (in von Jacobi's words) "to conceive the possibility that some motion or some mechanical operation might be produced by the electro-magnetic excitation of soft iron" (von Jacobi 1835). He had completed his first working motor in 1834, a little over a decade after Barlow's wheel made its first feeble rotations.

Moritz Hermann von Jacobi (Lithographer: Rudolph Hoffman; Photographer: Peter Geymeyer).

Along the way to the world's first practical electric motor there were many trials and even more errors—so many errors, in fact, that next time you turn on your hair dryer or cordless drill you might take a moment to reflect on the fortitude of Moritz Hermann von Jacobi, who, while he was at work in the laboratory, might well have received a generous hazardous duty allowance. Remember this: the handles of the tools in von Jacobi's workshop had no dipped vinyl, thermoplastic resins, or dual duro meter molding to shield their user from the jolts and surges of high voltage electric currents. The consequences of working with potent, live circuits using non-insulated tools were as dire as they were predictable. "During the motion of the apparatus," von Jacobi notes dryly at one point in his memorandum, "I received violent shocks and felt an extreme pricking sensation in the upper part of my body" (von Jacobi 1835).

History's first truly useful electric motor could have run something like a back yard string trimmer; it won its inventor a prize and an honorary doctorate from the University of Königsberg. With this initial success, von Jacobi was ever afterwards like a bloodhound on the hunt. Four years later, he built a much larger, second iteration of the same device. This time, he wasn't fooling around; the new motor was considerably more potent, to say the least. It weighed in at almost a quarter ton and could generate more than 300 watts—twenty times more powerful than his first. Weed-whackers be damned: old Moritz installed this behemoth in a twenty-eight-foot paddle boat, where for several months during the summer he ferried as many as a dozen passengers at a time back and forth across the Neva River, maintaining an average speed close to two miles an hour. Less than a year later, the intrepid captain retrofitted his vessel with a newer, even bigger motor producing a full thousand watts (1 kW) of power. At three to four times the power of the original motor, the boat's speed was just about doubled (Doppelbauer 2019).

Hindsight is always clearer, and one wonders how much cleaner and less cluttered the world might be if in or around 1904, the then nascent automotive industry had put more effort into outfitting horseless carriages with electric motors instead of inefficient, gasoline-burning, hard-to-start, reciprocating internal combustion engines. The single, fatal disadvantage of early electric cars seems to have been their relatively limited range. (That's still a problem with contemporary electric vehicles, of course, albeit less so.) But their benefits then were the same as they are now: electric vehicles were quieter, simpler (many fewer moving parts), and, because electric motors function without combustion, they produce no exhaust gases to offend noses and foul the atmosphere. And they've got two further advantages: an electric motor, unlike a piston engine, produces maximum torque as soon as you turn it on, and—even more important—it doesn't particularly care whether it runs fast or slow. Thus, an electric motor needs no gearbox, clutch, or shift lever.[2] But big or small, DC or AC, AAA battery-powered beard trimmer or a 240-volt, 30 amp vented clothes drier with steam and wi-fi connections, they all "make iron come."

CHAPTER 7

In the Sky

The Wright Stuff

The last day of the Air Show, 10 November 1912, the *Salon de la Locomotion Arienne* at the Grand Palais in Paris. Fernand Leger, Marcel Duchamp, and Constantin Brancusi stroll in silence among the exhibits. Balloons, dirigibles, the futuristic monoplane of Louis Blériot. Amid the motors and propellers, the men pause. Duchamp, who was later to win fame with his artistic "ready-mades," turns to Brancusi: "Painting has come to an end. Who can do anything better than this propeller. Can you?"

Duchamp's offhand remark must have stirred something in Brancusi's artistic soul; a decade and a half later, he will title one of his sculptures "Bird in Space." The piece is five feet high, slender and streamlined, cut from marble. At first Brancusi's sculpture is denied the status of artwork because (in the words of the New York *Journal-American*), it looks "like nothing so much as half of an airplane propeller" (Danto 1996). In time, however, Brancusi's work becomes part of one of the greatest assemblies of sculpture in modern art. The series of works (seven marble, nine bronze) are housed in museums in New York, Philadelphia, Seattle, Venice, Canberra; the original (1923) was sold in 2005 for $27.5 million.

Propellers are beautiful things. So are wings, which they resemble. The contemporary, conceptual artist Iñigo Manglano-Ovalle included both objects in a solo installation ("Happiness Is a State of Inertia") at the Monique Meloche Gallery in Chicago (Meloche 2013). Doubtless inspired by the Salon of Aviation display a century earlier, Manglano-Ovalle also included some propellers in his exhibit. A sculpture called "Untitled" consists of a pair of carved maple and aluminum blades. They're facsimiles of the propellers that had once attracted the attention of Duchamp and Brancusi; here they're propped on the floor against a wall in one section of the gallery.

The centerpiece of the exhibition is a work entitled "Drone Wing." It's a huge, white wing suspended at a slight angle from the ceiling. Modeled full size after the wing of an MQ-1 Predator, Manglano-Ovalle's creation was supposed to feel threatening if you stood near or under it. The artist seems to have intended a specific criticism of drone warfare, a message in keeping with the politics of much of the rest of his art. But most visitors seemed awed by the wing instead. They were beguiled by its aesthetics, its size, its splendor. It loomed majestically, sublimely overhead, something like the 92-foot blue whale that hangs from the ceiling in the National Museum of Natural History. The thing was so big, so white, so quiet, it was "more of a giant feather," says Jason Foumberg in *Frieze*, "than a lethal vehicle" (Foumberg 2013).

Before there were propellers there were airscrews (they're still called that in Britain) to distinguish them from the propulsion mechanisms (or "screws") on boats, and before that there were screws for fastening pieces of wood together, and before *that* there were screws for pressing grapes and lifting water out of dug wells and from along the banks of the river Nile. In what may have been another of his "Eureka" moments, Archimedes pictures a long, hollow tube inside which was fitted a snugly and upward-spiraling (or helical) shaft. When the shaft was rotated by means of a crank, the bottom surfaces scooped up water; as the device was spun round and round, the water was progressively raised higher and higher until finally it poured out the top of the tube. Large versions of Archimedes' screw still drain Dutch levees; even older models (according to the Greek historian and geographer Strabo) may have kept in bloom the fabled Hanging Gardens of Babylon.

The first screw to emerge from the water into the sunlight was installed on a hot air balloon belonging to Jean-François Blanchard. On October 16, 1784, Blanchard took off with a propeller made from three metal paddles attached to poles and a hub that could be spun rapidly by hand crank. The contraption couldn't generate nearly enough air flow to move the balloon along, but Blanchard persevered with his experiments for some years until he died from injuries suffered when he fell from his balloon; his widow carried on his ballooning demonstrations for some years until she, too, met the same fate.

To move forward, birds flap their wings, pulling themselves through the air as swimmers crawl through the water. Somewhat logically, the creators of early flying machines thought it was primarily the flapping that kept birds aloft, and so many of their experimental aircraft attempted to imitate the rapid, up-and-down motions of birds' wings. But these early attempts by humans to duplicate the mechanisms of nature led always to failure and occasionally to catastrophe. Flapping, says Stephen Vogel, "entails great mechanical complexity, not to mention a bunch of aerodynamic pitfalls" (Vogel 2004, xiv). Aircraft designers eventually learned that there were much more mechanically efficient ways than flapping to move aircraft through the skies.

When, in July of 2019, I rode along with my son piloting a Cessna 182, the wings were fixed solidly in place. They were responsible solely for keeping the plane aloft. All forward movement, however, was provided by a rapidly spinning propeller. At first glance, flapping wings and spinning propellers look nothing alike. Yet their functional similarities transcend any innate differences. The brothers Wilbur and Orville Wright of Dayton, Ohio, were among the first to comprehend that propellers and wings were sisters under the skin.

The Wright Brothers began their investigations into powered flight quite logically with a review of whatever literature on the subject existed at the time. But they soon discovered that marine screws of any sort were poor guides for modeling an aircraft propeller. "So far as we could learn," they wrote, "the exact action of the screw propeller, after a century of use, was still very obscure." More tests, more research. At last came the breakthrough: "[i]t was apparent," Orville notes drily in his journal, "that a propeller was simply an aeroplane [i.e., a wing] traveling in a spiral course" (Wright 1953). There was the answer the brothers had sought: their propulsive force would be nothing more than a wing flipped sideways, rapidly spinning! The twin, eight-foot spruce blades on the Wright Flyer were soon reconfigured to resemble the airfoils for wings that the brothers had been testing in their shop.

Here's how a wing works. Like water in a stream flowing round a rock, air flowing over a wing deforms to follow the contours on its upper and lower surfaces. The airflow will be slightly constricted and speed up as it moves over the curved region, and this condition tends to lower (according to Bernoulli's principle) the static pressure directly above it. The difference in pressure between curved and flat sides will be proportional to the differences between the squares of the respective speeds. A small difference in the speeds can produce a much larger difference between pressures, in effect "pulling" (or pushing) the wing in the direction of the faster flow. It doesn't matter whether air blows over the wing or the wing moves through the air. Either way, writes Doug McLean, a retired technical engineer from Boeing, "a diffuse cloud of low pressure always forms above the airfoil, and a diffuse cloud of high pressure usually forms below. Where these clouds touch the airfoil, they constitute the pressure difference that exerts lift on the airfoil" (McLean 2012).

But that's not the whole story, because there's some Newtonian mechanics mixed in that adds a second, upward force to a moving wing.

Replica of one of the propellers fitted on Orville and Wilbur Wrights' motor-driven airplane (1903) (Falcon Photography, 2015).

After a volume of air passes over the curved upper surface of a wing, it flows off the trailing edge directed downward at a slight angle. This volume of downward flowing air is called the "downwash," and, by Newton's third law (for every action there is an equal and opposite reaction), the downwash produces an upward force on the wing. The steeper the angle of downward deflection, the greater the downwash and the consequent opposite force that lifts the wing into the skies.[1] If you're a bird, all this buoyancy comes for free, or almost so—just for the price of flapping now and then to compensate for drag. Who would have thought it possible? Nature never gives away a free lunch, but, says Professor David Alexander, "getting lift from a wing is a pretty good bargain" (Alexander 2004, 35).

Turn a wing sideways, though, spin it like crazy, and you've got yourself a propeller that generates "lift" in a horizontal direction. Slice almost any wing and propeller crosswise and at right angles to their surfaces; you'll see right away the same thing as Orville and Wilbur. The top (or leading) side of each curves steeply up (or forward) and then tapers smoothly away and down to a sharp edge. The bottom (or trailing) surface, in contrast, is mostly flat or curved very slightly inward. Both wing and propeller work simply by moving those differentially contoured surfaces swiftly through the air. Air flowing from front to back over and around a wing produces a net upward force called "lift"; in the same way, air flowing across rapidly spinning propeller produces a net force to the rear called "thrust." The spinning blades want to move forward; the aircraft that's attached to them just comes along for the ride.[2]

The faster a wing moves, the more lift it's capable of producing; likewise, the faster a propeller spins, the more air it can shove backwards. But faster-spinning propellers are not always better. The outer parts of spinning propeller blades are moving at much greater velocities than the inner parts, and as the tips of the blades approach supersonic speed (remember Benjamin Robins' experiments?) they become much less effective at moving air. There was a time in the middle of the twentieth century when aircraft designers actively pursued supersonic propeller design, but those efforts were largely abandoned when it was discovered that blades' aerodynamic efficiency fell off seriously when the tips exceeded the speed of sound. In one extreme case, a developmental program was dropped after workers became nauseous during ground tests of supersonic propellers that generated air waves ("sounds") that could not be heard.

The pressure fields that form on opposite sides of a moving wing or spinning propeller can be produced when air or other fluids flow around any curved, rotating surface such as a sphere or a cylinder. As a result of his experiments with streams of fluids, Gustave Magnus had guessed that the force produced by cylinders spinning in a fluid medium was essentially the same as the "lift" produced when a wing moves through the air. He was right: spinning tubes, just like wings, can generate forces sufficient to move them in one way or another, or even, if they spin fast enough, to keep themselves buoyant.

As the nineteenth century ended, word about Magnus's discoveries had permeated the engineering community, and when small, self-contained internal combustion engines became widely available in the early 1900s, inventors began to sketch plans for Magnus-principled flying machines. One of these dreamers was Spanish-American war veteran Butler Ames. In 1908 Ames, by then a congressman, commissioned the building of his Aerocycle following the basic mechanical principles outlined by Magnus. Like the Wright Flyer, Ames' machine married solid aerodynamic theory with a generous measure of intrepidity.

The Aerocycle (for which Ames later obtained a patent) was supposed to work like this: a pair of large, lightweight cylinders, mounted horizontally, would be set spinning furiously, and the lift they generated would theoretically be sufficient to raise themselves and everything else attached to them into the air. Ames used a gasoline V-8 engine to spin both cylinders as well as to turn a small, pusher-type propeller so the machine, once airborne, would be able to move forward. Whether the Aerocycle ever "flew" (let alone flew well) is a matter of some debate. An entry in the article in the *Army and Navy Journal* dated July 30, 1910, reports that Congressman Ames had "tested" the Aerocycle aboard the

USS *Bagley*, and a contemporary report in the *New York Times* testified that the machine rose a little distance off its platform (Hoppe 2021).

A century later, the sky is not exactly full of Magnus-force-powered aircraft. But the physics behind them is sound, and experiments with using spinning cylinders as sources of aerodynamic force continued for more than a century after Ames' brainchild took to the air (or didn't). Spinning cylinders are fully capable of keeping very small flying machines aloft, in any case. I remember a conversation with my dentist, Benjamin Bowen. When he isn't repairing teeth, Bowen dabbles both in toy airplanes and the ballistics of hunting rifles. In his office that morning, staring up at a half dozen toy propeller-driven airplanes suspended on strings from the ceiling, I mentioned that I was writing a book on things that spin.

Bowen immediately put down his drill, and, as we waited for the painkilling drug to take effect, he swiftly and enthusiastically launched into a detailed explanation of the Magnus effect. Since I could feel neither lips nor lower jaw, I couldn't add much to the conversation. But weeks later Bowen and I had an actual dialogue about Herr Magnus. He started with a demonstration: "You can actually make a Bic pen fly," he said, grabbing one from the cup of sanitized pens on the counter. Holding the plastic tube aloft horizontally, Bowen snapped his fingers and spun the pen off his fingertips. It zipped across the room, holding plane, to my amazement, just like Graf's underspin backhand; I would have sworn the thing was floating.

Rotating, lift-producing cylinders are most often called Flettner rotors—after yet another German aerodynamic scientist and inventor, Anton Flettner. A Flettner rotor can produce energy much greater than you'd expect from a laboratory curiosity. In 1926, for example, Flettner stood twin 50-foot-high cylinders straight up on the deck of the German ship *Buckau* and used the horizontal "lift" they generated when spinning to power the boat all the way from Europe first to South America and then on to New York.

Elmer's Song

George Albert Tuttle, my mother's cousin, was graduated from the United States Military Academy on June 4, 1944—the same day, in one of those quirky overlaps of history, that my wife was born. Lieutenant George Tuttle survived the Battle of the Bulge late that year. After the war, Tut and his family became frequent visitors to my parents' household. I still have his West Point yearbook, *The Howitzer*; he had given the book to my mother, and it passed on to me after her death. Sometimes I take the olive-drab book off the shelf and browse through it. Scanning the faces, I wonder who among them never lived to see another June. But the most engaging part of the yearbook is not the pictures and bios of 470 newly commissioned officers, but the hundred-odd pages of advertisements clustered at the back. The pages swirl with the names of long-vanished products and once distinctly American corporations: Hallicrafters, Martin Aircraft, Erie City Iron Works, the Bendix Corporation. The ads play like a soundtrack from long ago. Flipping through them I hear the static of old radios, the hiss escaping from a six-ounce Coke bottle, the throbbing drone of a squadron of propeller-driven aircraft passing overhead.

Page 541 catches my eye. It's an ad for the Sperry Gyroscope Company of Brooklyn, a six-stanza poem entitled "Song of Elmer … the pilot who never gets tired." The verse is doggerel, the sentiments banal. It's been a long time since advertisers fed poetry to customers. But the description of a wondrous gyroscopic device known as an inertial guidance system is dead-on:

> He can hold a plane on a chosen course
>   while the crewmen rest or sleep,
> He can level off for a landing glide
>   Or bank her sharp and steep—
> He can spiral up, he can spiral down,
>   Or hold her level and true—
> His hydraulic muscles never tire
>   The way human muscles do....

The Sperry Corporation no longer exists; it prospered during World War II, built computers in mid-century, then broke into smaller manufacturing divisions bought by companies like Honeywell, Ford, and Northrop Grumman for whom "Sperry Marine" in England still builds navigation and communication equipment. But nearly identical gyroscopes still point the way for the Hubble Space Telescope, installed in six separate rate sensor assemblies. These six-pound, electrically spun devices have kept Hubble aimed in the right direction for the last thirty years. (As I write, only three of them are functioning.) The diminutive gyroscopes (3 × 6 inches each) aren't a whole lot bigger than the toy I played with as a child. Yet they aim the twelve-ton, school bus–sized telescope at the stars with an accuracy, according to NASA, equivalent to spotting a laser beam on the head of a Roosevelt dime from two hundred miles.

When, in 1974, my friend Eino Jacobson quit homesteading in the back country of north Idaho to try his luck as a carpenter working on

Alaska's North Slope, he hitched a ride out of the Spokane Airport with a friend flying a Super Cub headed north. The navigational plan was simplicity itself, a resurrection of the days of barnstorming: fly low and follow the road. (In this case, the road for most of the trip was the Alaska Highway, which even then ran from the northern border of the state of Washington all the way to Whitehorse.) Modern air travel is no longer such a hit-and-miss business.

Years later, driving to the same airport in October of 2018 to meet Orville Pierson, an old friend from college, I tried to picture him and me speeding along different vectors, seeing different things, simultaneously arriving at a common place. I was still dependent on following a road. But modern aircraft have no such limitations; they keep track of their location by means of several electronic media, including signals flashed from beacons to planes (ADF and VOR systems), radar, and global navigation satellite systems (GNSS). So much geo-positioning redundancy is available to pilots and air traffic controllers these days that so long as a plane remains in the air somewhere, anywhere at all, it's almost impossible for a pilot to become lost.

Each of these newer navigational systems depends on some interaction between the plane and things under or over it. The advantage of older inertial navigation systems (INS), on the other hand, like that advertised in 1944 by the Sperry Company, is that a pilot can fly blind, so to speak. The "Song of Elmer" nails it: with Elmer (or his progeny) in the cockpit, human pilots can tell whether they're climbing or descending, banking left or right, any time of the day or night, in bright sunshine or through thick clouds, just by listening to what Elmer tells them.

The first long-distance navigators who had to keep track of their location on featureless (or nearly featureless) domains were sailors. Getting from one place to another and back again is relatively easy when there are landmarks to follow. But what do you do when your destination is out of sight to begin with, when all you know is that it's somewhere *out there*? The easiest and safest course for early sailors to navigate was to hew to known features on the seacoast. For centuries, this was how Greek and Arab sailors circumnavigated between ports of call on the Mediterranean or ventured eastward across the Indian Ocean. Odysseus, the legendary voyager of ancient Greek myth, would probably have plied the Mediterranean Sea by keeping within sight of a coastline or sailing toward recognizable topographic features.

Voyaging across open waters was vastly more challenging and hazardous. Far from land, you mark your place according to its latitude and longitude, a graph-paper grid laid everywhere over the globe. The lines of latitude run east and west; they define, in turn, your relative position north or south of the Equator, whether nearer to the Equator or closer to the poles, as in: 20 degrees North or 80 degrees south. On the open ocean, to determine your latitude is relatively easy. All you must do is measure the angle of the sun at the meridian, then compare it to known values for that date. If the noonday sun happens to be directly overhead on the vernal and autumnal equinoxes, for example, you know you're somewhere on the Tropics of Cancer or Capricorn.

Determining your longitude—more specifically, determining your divergence from a starting point when traveling in a westerly or easterly directions—is more problematic. To know your longitude when the sun has climbed to the meridian you must include the current time *at your home port* into the calculation. Precisely *when* the sun reaches its meridian varies with one's location on an east-west circumference; when Big Ben tolls at mid-day, the citizens of New York are cooking breakfast, while in Los Angeles most folks are still fast asleep. Only by comparing *your* noontime with noontime at the point of your departure can you tell how far west or east you have traveled. At the equator, for example, every hour's difference between you and your place of origin equals a thousand miles of distance; close to the poles, however, that hour's discrepancy would scarcely account for more than a snowball's throw.

Aircraft pilots faced this problem too, and in the early days of long-distance air travel pilots commonly relied on the same way-finding methods used by sailors of old, keeping track of their location relative to the surface of the globe by using clock, sextant, and stars. Pilots also had a new navigational problem, however, because an aircraft moves not just across a two-dimensional surface, whether land or water, but through *three*. This required a different kind of navigational instrument to tell whether an aircraft was climbing away from the earth or toward it, banking left or right; says veteran pilot of the Boeing 747, Mark Vanhoenacker: "The attitude of a plane—the angle of its nose in the sky—is so critical to flight that [an attitude indicator] dominates the central screen, the primary light display, in front of each pilot" (Vanhoenacker 2015, 80). The instrument pilots adopted for this purpose was a gyroscope—to be sure, it was a fancy, delicate, and expensive gyroscope—but it worked the same as any child's plaything. By using the principle of the conservation of angular momentum, the gyroscopic compass informed pilots which was left and right, up and down.

The earliest gyroscopic instruments installed in airplanes were purely mechanical attitude-indicating devices; they kept track neither of velocity nor location. "Elmer" carried information in the form of an initial quantity of angular momentum, which he conserved as zealously as a miser his hoard. Set spinning just before takeoff, a gyroscope automatically retained its initial angular momentum, and so the instrument was instantly able to inform pilots when the aircraft was rolling left or right, pitching up or down. The poet "Elmer" spoke the truth: in sunshine and in the darkest night, hurtling through clouds or driving snow, the gyroscope was an inerrant, indefatigable co-pilot. Tune the plane's gyroscope with respect to a particular directional heading, and pilots and crew, as the wartime poem says, can rest or sleep.

Even the most bare-boned airplane cockpit of today will have at least three gyroscopic instruments. These include a turn-and-bank indicator to detect yaw (left or right movements of the aircraft); a gyro-compass or directional gyroscope, set initially to point north and to detect any deviation from the plane's desired course heading; and an attitude-indicator that displays an artificial horizon to show changes in pitch (nose up or nose down). Modern gyroscopes use light-based technology rather than spinning discs or wheels, but a ring-laser gyroscope, as it's called, works basically the same way as one plucked from a toy-store shelf.

Here's a simplistic explanation. A beam of light is split at its source and sent around the instrument (sometimes called a ring interferometer) on the same path but in opposite directions. When the beams return to the starting point they are measured to see if their respective wave patterns overlap or interfere with one another. If both wave frequencies coincide, that means that there have been no changes in the orientation of the apparatus, absolute proof that the aircraft hasn't rolled or pitched. If, however, the waves are seen to be out of step, this indicates that one of the beams has had to travel further than the other because the angular velocity of the interferometer has been changed.

The ring-laser gyroscope is a fantastically ingenious device; its chief advantage over mechanical navigational instruments like Elmer is that a ring-laser gyroscope possesses no moving parts to wear out. Just like Elmer, however, the ring-laser gyroscope doesn't need any source of outside information whatsoever to tell pilots whether they're going up or down, left or right. It's uncanny; the thing somehow *knows*, says Vanhoenacker, "without looking at stars, maps, satellites, or scenery, without interrogating anyone or anything" (Vanhoenacker 2015, 81).

Aero-Tops and Puddle Jumpers

Long before there were Wright Flyers and MQ-1 Predators there were toys. My son fiddled with small propeller-driven airplanes in high school as part of an annual interscholastic competition known as Science Olympiad. The program dates back almost half a century when several of the faculty at Andrews Presbyterian College in North Carolina designed an academic competition for regional high school students in the sciences—originally, biology, chemistry, and physics. The purpose was to popularize science by making it both intellectually interesting and fun; early competitions, for example, included events featuring paper airplane design or "beaker races" in which teams competed to see who could fabricate an insulated container that would keep a beaker of water hot for the longest amount of time.

One event was called "The Wright Stuff." The object was to construct a small, rubber-powered model airplane and to fly it for as long as possible. Shape and overall weight were important considerations. But it was the propeller that was key to success or failure. Students' planes were restricted to commercially available, fixed diameter, injection-molded plastic propellers. Competitors could scrape or sand the blades to make them lighter but not otherwise reconfigure them. With meticulous trimming and balancing, these stubby red blades could be shaved thin to a translucent pink, spinning on a knotted loop of rubber for as long as three or more minutes.

How new and how old such technology seems! Leonardo da Vinci sketched numerous designs for flying machines powered by aerial screws. Three centuries later, Mikhail Lomonosov used the mainspring of a clock to spin a pair of contra-rotating propellers attached to a small box in which he hoped to lift instruments into the clouds to conduct meteorological experiments. Lomonosov called his flying machine the Aerodynamic; he arranged a demonstration before the Russian Academy of Sciences in July of 1754, but the "Aerodynamic" merely dangled from a string.

In 1784, Christian de Launoy, a French naturalist, and his mechanic (who went by the name Bienvenu), built a more credible propeller-driven toy aircraft. Not much is known about Launoy, but his sketch of the device still exists. The drawing shows a stick with four turkey feathers attached to a cork hub at each end; a slender bow and bowstring are also centered on the stick between the feathered ends. As the bowstring was twisted round and round the stick, it put increasing tension on the bow; when released, the bow spun rapidly, releasing stored energy by turning both sets of feathers. The spinning of the feathers, according to Launoy's notes, caused "vertical percussions" sufficient to raise the machine.

There's no evidence that Launoy's flying machine ever took to the air. But unlike Leonardo's or Lomonosov's models, it *could* have. In 1975, a San Francisco company called Aero-Motion developed a similar toy called "puddle jumper," after the slangy, affectionate term used to describe small, single-engine short-haul passenger aircraft. Aero-Motion's toy consisted of a pair of lightweight, hardwood blades attached radially to a pencil-sized stem; when kids (and adults, too!) placed the stem between their palms and spun it rapidly, the spinning blades created a sudden pulse of aerodynamic thrust that made the toy jump into the air like a flushed quail.

Kids still play with these low-tech throwbacks to an older, pre-electronic childhood; a box of six Puddle-Jumpers sells on the company's website for about $25, but cheaper, ultra-lightweight plastic versions are also widely available in big box toy aisles and discount drug stores. These appeal as much to adults as to children; a few years ago, Jennifer Davis, office manager of the Colorado Flight Center, a flight school at the Grand Junction Regional Airport, blogged that the tiny, whirring copters were "a big hit with our flight instructors" (Davis 2019). Puddle-Jumpers' new "Aero-Top" is a larger, plastic version of an original wooden "finger top." Both wood and plastic tops are said to keep their users "spinnin'" and "grinnin'," and both come with detailed explanations of the physics of rotational energy, friction, and gravity.

Toy propellers have a long and distinguished pedigree. They may have originated long ago in China in the fourth century BCE when children first fastened bird feathers to small sticks, twirled them rapidly between the palms of their hands and then let go. Were the children practicing sympathetic magic by attaching pieces of life from the sky to their tiny, earthbound sticks? Or did they see more deeply into the essence of propeller-driven flight? Did they know that the rapid movement of air over the top of a gently curved surface such as the upper side of a bird's wing feather produced a mysterious, upward force? From these crude toys to the functioning toy helicopter built in 1796 by the British scientist George Cayley, to the twin, laminated spruce propellers Orville and Wilbur fitted between the wings of the Wright Flyer, to the eight-blade, counter-rotating, scimitar-shaped, carbon-fiber-reinforced propellers mounted on the wings of the 150-ton Airbus A400M military transport, the family resemblance stands out.

Leavin' on a Jet Plane

Jet aircraft may be missing propellers, but they do not lack spinning blades that move large volumes of air. In fact, a modern turbofan engine like the GEnx (for "General Electric next generation"), developed specifically for the Boeing 787 "Dreamliner," contains eighteen air-moving blades. You just can't see them. The reason the spinning blades on a jet engine are tucked away inside a large cylindrical shroud, rather than mounted up front where it's obvious what they're doing, is not that they're performing a different function. Those whirling blades have the same purpose as the propellers on the Wright Flyer, to move large amounts of air as rapidly as possible.

When you watch a jet plane accelerate down the runway and lift itself into the air, it's not obvious that dozens of spinning blades are driving the aircraft forward—indirectly, at least. Allied pilots who first encountered jet attack aircraft near the end of World War II were stunned as much by the planes' hidden source of power as by their amazing speed: "The bloody thing's got no props," radioed one pilot of an RAF Mosquito that came under fire from a German ME 262, the world's first operational jet-powered fighter aircraft. Jets like the ME 262 are propelled forward by an invisible, high-pressure stream or "jet" of exhaust gases blasted violently from a nozzle facing to the rear of the aircraft. The force of the outward flowing stream causes an equal (but opposite) force against the plane, shoving it vigorously in the opposite direction. At any given moment in the day, according to data tracking companies like FlightAware, almost 10,000 jet aircraft are in the sky proving Newton's third law of motion.

Yet jet engines spin. Anyway, critical parts of them do. Sometimes when you walk down the jetway to board an aircraft you can glimpse the engines, hanging beneath the wings. Inside the cowl you see a large, fan-type wheel with many stubby blades. Usually, the blades are stationary while the jet is parked for boarding, but sometimes you can catch them rotating slowly and soundlessly. The engines on commercial aircraft often have a white, spiral design painted on the cone at the center. It looks cool, and if the engine is spinning slowly the effect is mesmerizing, mildly hypnotic. But the spiral isn't there for show. The intake openings on a passenger jet are big enough to swallow a person, chew them up. The artsy design is put there to alert ground crews when the engine is running; if the spiral is moving fast enough to blur, workers know to keep a safe distance away. A jet engine is beautiful and dangerous simultaneously. Giving his 747 a pre-flight check, pilot Mark Vanhoenacker says that he sometimes gives its engines' fans "a casual, affectionate whirl." It's difficult to believe, Vanhoenacker says, "how easily such a vast wheel can turn" (Vanhoenacker 2015, 120).

Much like a propeller, the fan blades on a jet engine shove huge volumes of air to the rear. But the similarity with propeller-driven aircraft ends there. Immediately behind the big, outer fan inside a jet engine are rows of smaller fans stacked one after another like slices of zucchini on a skewer. These spinning "compressor" fan blades boost internal pressure within the overall air flow through the engine; each successive set of blades increases the air pressure slightly so that by the time it's ignited inside the combustion chamber the air molecules are jammed together several times tighter than when they were first sucked into the intake.

Once ignited, this compressed air makes for a more energetic burn. Exhaust gases from combustion then flow through multiple sets of airfoil "cascades." These rows of fixed blades keep the gases flowing parallel to the longitudinal axis of the engine, directing their force to the rear. Just before they exit, the gases pass through one final set of spinning blades. These fans absorb a small quantity of energy from the exhaust flow and direct it back to the compressor at the front of the engine, keeping the whole compression/combustion/exhaust process running. But most of the outflow gas is necked through a tapering exhaust nozzle, increasing its velocity—think of squeezing your thumb partly across the flow of water from a garden hose—and driving the plane forward.

Moulin Rouge

Join two paddles opposite each other at a hub. Lift the device you have made into a strong breeze. Feel the wind blow against the blades, bow them backward, try to rip them from your grasp. You think: *they contain the power of the wind*. Now, cock the blades slightly to vector that force to the side. Feel the blades twist, torque against your wrist. Something in your brain catches life, blooms: you sketch your creation in your journal; this thing, you realize, can capture the wind to turn a shaft, lift a piston, play a tune.

"Of all natural phenomena," said the renowned anthropologist James George Frazer, "there are, perhaps, none which civilized man feels himself more powerless to influence than the wind" (Frazer 1894, 19). Early attempts to master the wind were more often combative than practical. In the uppermost latitudes of North America, says Lyall Watson, where "the wind blew for weeks on end and the Inuit began to despair of ever going out hunting again, they made a long whip out of seaweed and struck out in the direction of the gale, crying '*Taba*! It is enough!'" (Watson 1984, 116). When that didn't work (as was, presumably, often the case), the tribe took a more aggressive approach by assembling round a fire, pissing communally in a large jar, then hurling the pot into the flames, on the assumption (in Watson's description) that "no self-respecting demon would want to hang around after such an insult" (Watson 116).

Similarly belligerent tactics were used to control the wind says Watson, by the Xhosa of South Africa who spat a specially brewed noxious potion into the eye of the gale, by the Kai people of New Guinea who lashed spears to the rooftops to pierce the wind's belly, or by the Scottish highlanders who hurled shoes (though for some reason only their left ones) into the teeth of the blast.

In other places, people struck a more conciliatory tone, attempting to beguile the fickle winds with charms, potions, and incantations. Farmers in ancient Persia tossed handfuls of powdered saffron into the air in the belief that the drifting, gold-colored clouds would bring on the sustained, gentle winds necessary for winnowing grain. The Berbers of North Africa, on the other hand, stuck flags in their grain piles, hoping that the pennants' desultory fluttering would catch the wind's eye and prompt it to blow with somewhat greater diligence.

Sailors in particular—whose livelihoods if not their lives depended on favorable winds—went to extraordinary lengths to curry the winds' favor. On Murray Island off the northernmost tip of Australia's Queensland, writes Watson, the way to get a favorable wind for sailing was to arrange the leaves of a coconut tree to point in the desired direction. For mariners attempting to negotiate the variable currents around Vancouver Island in the Pacific Northwest, however, the trick was to take one of a permanent set of "wind stones" and reposition it ever so slightly to coax the wind to blow briefly in just the right direction. Those in need of wind often made use of a little sympathetic magic, whistling softly for a light breeze (say, for winnowing grain), somewhat louder if stiffer gusts were called for.

The earliest recorded account of tapping the power of the wind to spin fan-shaped blades on a central shaft comes from *Pneumatica*, a journal kept by Heron of Alexandria (first century CE; Shepherd 1990, 6). Heron describes an organ being supplied with air for its pipes by a wheel which has oar-like scoops as in ancient "wind vanes" (*anemourion*). Though contemporary testimony is lacking, there's no question the thing described by Heron would have worked, perhaps must have worked. Detailed accounts of an actual windmill used to process grain and pump water, however, did not come for more than a thousand years later in Sistan in eastern Persia near the border of Afghanistan. There, in "the land of wind and sand," wrote Al-Istakhri (c. 950 CE), "the wind drives mills and raises water from the streams, whereby gardens are irrigated" (Shepherd 1990, 8).

Those first windmills didn't look much like "windmills" at all. Instead of spinning on a horizontal axis like a giant pinwheel, they spun round a vertical axis like a merry-go-round. Reconstructions of these early windmills resemble a circle of dominoes stood on end around the perimeter of a circular platform, or the small, slotted carousels once used to house photographic slides. Fabric panels, stood upright in a circle, caught the wind and turned a central shaft. A hood installed on one side of the mill blocked the panels moving upwind from being blown backward. The mechanism turned slowly, speeding up and slowing down with the fitful breezes.

These horizontal mills (or *panemones*, "things of all winds") were used at first to pump water, later to grind the caliphs' grain. By the end of the millennium, they had spread to India, China, Europe; the mills, writes Watson, "seems to represent a happy fusion of eastern and western ideas. The milling mechanism was copied from the Roman waterwheel, but [the panemones] were pure Tibetan prayer wheel" (Watson 1984, 121). Compared to stream-powered mills, panemones were inefficient. Yet they had two inherent advantages over hydropower: the wind blew during winter months when rivers were frozen, and it blew also inside the walls of besieged towns.

The iconic, horizontal-axis windmills of popular imagination were not part of European landscapes until the late middle Ages. The earliest reference to such mills comes from a twelfth-century manuscript which refers to a certain plot of land "near a windmill." First-generation European windmills were built atop a massive, underground post; tons of wood, brick, and stone had to be nudged round almost daily so that the entire structure with its blades ("sails") could face into the prevailing wind. By the end of the thirteenth century, mills were engineered more sensibly so that only the uppermost "cap" which housed the blades and central shaft had to be rotated. Blades typically were constructed like a set of Venetian blinds, each containing dozens of individual vanes that could be set at varying angles so that the mill spun efficiently and safely in virtually any wind. As many as 100,000 such machines were in operation throughout Europe until late in the nineteenth century.

An irony to contemplate: the most famous windmill in history never once spun its sails, never once ground so much as an ounce of flour. It was just a mill pretending to be one for old times' sake. The cabaret with the red windmill on the roof opened its doors to Parisians on October 6, 1889. And here's something else: why would you put a windmill on the roof of your nightclub in the first place? Stories about the creation of Moulin Rouge proliferate. Here's one: the district of Montmartre was once open French countryside, spotted with windmills that for centuries had been spinning to separate bran from flour, to crush spices and seeds. The iconic, red windmill preserves that time and place, like a keepsake. Another: the cabaret was built in memory of three brothers who were killed defending their own family mill against invading Cossacks, its crimson meant to commemorate the blood they shed. Yet a third story gives a nod and a wink to the red-light history of this section of Paris. From the sixteenth century on, the more than forty-odd windmills of Montmartre ground grain by day, but come sundown were repurposed as brothels.

The Door Swings Both Ways

Electric motors use a flow (or current) of electrons to create interacting magnetic fields that produce mechanical force, or torque. But the equation is reversible: while continental inventors were hard at work designing motors that *consumed* electrons, Michael Faraday at the Royal Institute of Great Britain had discovered how to *manufacture* the current that would keep those motors fat and happy. In October of 1831, he wrapped a length of copper wire around a tube and attached both ends not to a source of electric power, but instead to a galvanometer (basically a needle sensitive to the magnetic field produced by a live electric current). Then he took a long, slim bar magnet and poked it back and forth inside the tube of coiled wire. Immediately the needle on the galvanometer began to swing, indicating that an electric current was flowing through the wires around the tube. Had Faraday somehow been able to connect his apparatus to a flashlight bulb instead of a galvanometer, he could have seen the future light up before his eyes. His first experimental generator used linear motion to induce the flow of electrons, but it took no time at all for him to see that *spinning* the internal magnetic field rather than sliding it back and forth would be far more efficient and productive.

An electric *motor* needs a constant supply of electrons to keep spinning, whether from batteries or the outlets in a wall. But an electric *generator*, on the other hand, only needs some way to make it spin round and round. A generator doesn't care where the power to spin its armature comes from. Early commercial power stations were quite naturally built alongside flowing streams for the same reason that sawmills had been put there in previous centuries. The kinetic energy of flowing water had long been used to turn wheels and cranks; it could just as well spin electric generators.

Falling water possesses even more potential energy than a flowing current, and so hydroelectric plants were soon installed in dams on rivers where there was a significant drop in elevation over a relatively short distance. Water backed up behind a dam possesses a huge amount of potential kinetic energy. When it is released through sluice gates near the top of a modern hydroelectric dam, it plunges with great velocity through thirty-foot penstocks and into the blades of multiple large turbines, spinning generators capable of putting out a continuous 700 megawatts, an awe-inspiring million horsepower. *Touch it*, a guide from the Bureau of Reclamation once said to Joan Didion when she stood before one such turbine deep inside Hoover Dam. *And I did*, she said, *and for a long time I just stood there with my hands on the turbine* (Didion 1979, 200).

Big or small, it's all about making iron come. Electrical energy and magnets are inseparable; where there's one, there's the other. Available now are micro- and pico-sized hydropower installations that dribble out electrons sufficient for a couple of light bulbs or a small-screen television. At a ranch along the St. Maries River in the panhandle of Idaho, I tour a homestead-sized power plant. It's nothing like the large, slow-wheeling water mills of old. A six-inch pipe diverts part of the flow from a small creek, channels it along a hundred-foot drop and into a quadrant of nozzles spaced around the circumference of a turbine, then spills the outflow back into the creek downstream. Nothing is wasted, nothing lost. The nozzles are positioned about an inch off center to provide torque to spin the cast bronze turbine blades; the turbine itself (a nineteenth-century French word coined from the Latin *turbo*, or "spinning top") would fit easily inside a gallon can of paint, yet it sends thousands of watts to the house—hour upon hour, day and night, rain or shine.

The twentieth century witnessed great hydro-power projects like America's Hoover and Grand Coulee Dams, Egypt's Aswan, or China's Three Gorges. Now, it's just as likely the electrons on tap in your wall sockets are flowing courtesy of the wind. At the top of the scale of spinning propellers are the giant blades of wind turbines. I first saw wind turbines in 1984 in western Montana just outside the town of Livingston, Montana, on the Yellowstone River.[3] Each summer for more than twenty years my family and I drove west from our home in Atlanta to spend several weeks at the log cabin in the Idaho panhandle where we had lived while I was earning a PhD in English at Washington State University in Pullman.

We were never able to spend as much time there as we desired. On the one hand, we couldn't afford four (eventually five) separate air fares; travel by automobile, on the other, seriously ate away precious days. Most years we made the trip in several marathon, seven-hundred-plus-mile segments. Until the last sections of I-90 were completed at some time during the 1990s, we drove most of the miles through Montana on two lane blacktop, mostly US 10. The third day out of Atlanta always brought us at sundown to the Yellowstone River and a cluster of motels and service stations located east of town. Just before the intersection of US 89 and what was destined to become Exit 333 of I-90, we crested a long, gentle rise and came into view of three large wind turbines. Catching sight of them was both an arrival and omen; elegant and yet somehow menacing, they seemed built to some unknown, sinister purpose, like the engines in H.G. Wells' story of a time machine.

Wind turbine blades are 120 feet long; they're made from fiberglass and balsa wood, bonded together by epoxy resins and carbon fiber reinforcements.[4] Three blades sit atop a 260-foot tower, so that the overall height from the tower base to the topmost point of a spinning blade exceeds 380 feet. Looking at an array of such machines spaced evenly across a distant slope, it's easy to misinterpret what you're seeing. But then you walk amidst them for the first time, and it's not like what you expect at all. *Like trees*, you think. *Like a forest of giant white trees.* You're not just dwarfed; you're awed. The space feels both spooky and sacred. You think: *can these possibly be the same thing as the propellers on my son's toy airplanes? No, they cannot.*

CHAPTER 8

A Descent into the Maelstrom

Of Ice Disks, Toilet Bowls, and Edgar Allan Poe

Early on the morning of January 14, 2019, people strolling the shores of the Presumpscot River in the town of Westbrook, Maine, were surprised to see a gigantic wheel of ice revolving lazily counterclockwise in the current. The massive, slow-turning disk became an immediate celebrity—citizens congregated on the shore to watch the wheel, videos of it flooded most social media sites, and even the local ducks squatted down on their new carousel to preen and gawk at one another. The Associated Press picked up on the story, which is where I first learned about the event, and shortly afterward *Time Magazine* ran an online feature essay on ice circles and the physics behind them (Kluger 2019).

This extensive media coverage might suggest that the Presumpscot ice wheel was a rare phenomenon, but such ice disks form often in northern rivers. What made this wheel so interesting was its extraordinary size—at 300 feet across, it was more than ten times larger than is common. There's no reason why this one should have become so much bigger than most others. Like the storm in the Texas panhandle initiated by that legendary butterfly flapping its wings in China, this ice wheel, like all its smaller brethren, was born of pure chance. Random eddies in the river' current trapped small bits of ice, and over the course of several hours these slower, circling waters collected more and more pieces of ice until eventually they clumped to form a large, circular, slowly revolving sheet. Also contributing to the formation of this or any other ice wheel are some basic principles of thermodynamics. Any partial melting of the disk—as might happen, for example, when the winter sun falls on it in the late afternoon—takes place at its edges, and the water released from that melting tends to sink in spirals under the ice, introducing even more rotational energy to the system.

I first read about the Presumpscot ice disk in the *Moscow-Pullman Daily News* on the morning of January 16, 2019. Reading it triggered thoughts of cold northern landscapes and black, unforgiving waters, and then in an instant I was back in tenth grade and reading a story by Edgar Allan Poe called "A Descent into the Maelstrom." The first two centuries of Americans' literary history that year had been slow going. We had read, as I now remember it, some doggerel by Anne Bradstreet and Edward Taylor along with stale prose excerpted from the sermons of Jonathan Edwards and Perry Miller's *Errand into the Wilderness*.

All that changed when I came across Poe's story. The tale begins when two hikers reach the summit of a mountain overlooking the Norwegian Sea and numerous small islands along the coast. As the men pause on the cliff to take in the view, one of them, a former mariner, recounts the time his ship had been sucked into the Moskoe-strom, a great and dangerous oceanic whirlpool that forms daily amidst several of the islands visible now in the distance. It is, the narrator says, almost to a day the third anniversary of the awful event that took the life of his brother and very nearly his own. He himself had been spared from being swept down with the ship only by lashing himself to a large barrel and leaping into the sea. Because cylindrical shapes were being pulled downwards into the vortex more slowly than the larger and heavier ship, he was able to avoid being smashed against the rocky bottom until at last the maelstrom dissolved and he was buoyed back to the ocean's surface.

I had been curious whether Poe was making things up, so I did some research on oceanic currents and learned that the so-called Moskenstraumen was not another gothic fantasy. Nature, it seemed, really did harbor dire existential threats like this one. Firsthand accounts of this exact same whirlpool could be found throughout Norse literature and history, some written as early as the thirteenth century. Even the name Poe had chosen was real: *maelstrom* (from Dutch *maalen*, to whirl or grind, and *stroom*, stream) had been used by chroniclers in the 1680s to describe the sinister, swirling waters of the Moskenstraumen. Poe had simply introduced the word into the common English lexicon.

Nothing stood between me and the story this old seaman was telling. I could feel his dread as his ship was being pulled inexorably toward the giant whirlpool, a "terrific funnel ... smooth, shining, jet-black, wall of water, inclined to the horizon at an angle of some forty-five degrees." I imagined clearly what it would be like to be trapped deep inside such a vortex, looking skyward toward the surface of the ocean that "towered above us, a high, black, mountainous ridge." Poe may have exaggerated the size and intensity of nature's Moskenstraumen for dramatic effect, but the terrors of "A Descent into the Maelstrom" were real enough to haunt me for days.

Something there is that doesn't love a vortex, and that something is us. Poe's story isn't the only imaginative account of death by disappearance down a spinning hole. It's no accident that one of the most common circumlocutions for dying is "circling the drain." A vortex doesn't even have to be particularly large to have this spooky effect. The fear of being sucked, swirling into oblivion seems basic, almost inborn. When she was two years old, my daughter Laura was adamant in refusing potty training. Her behavior wasn't borne out of stubbornness. Rather, as she was able to articulate more fully a year or so later, the problem was with the flushing. She was afraid to sit atop the toilet seat for fear she'd be accidentally swept down the bowl. The solution—don't ask me why it worked—was to install a bottle of Ty-D-Bol in the tank reservoir. Somehow the deep, lapis blue color erased whatever threat was posed by the swirling waters as they disappeared down the hole.

Laura wasn't alone in her fear of being sucked out of existence; the phobia is common enough for Hollywood to exploit. In the popular 1996 film *Twister* (a movie that still shows up regularly on streaming channels), one of the central characters, the meteorologist Jo Thornton, suffers from the condition known as lilapsophobia because as a child she witnessed her father die in just such a way. (The name comes from ancient Greek *lailaps*, a whirling, tempestuous wind.) Fears of vanishing down (or up) a vortex extend even to the galactic scale: in 1979, Walt Disney Productions bet on people's anxieties with their film *The Black Hole*. This was a futuristic "space opera" about an American spacecraft's encounter with a supermassive gravitational field, a swirling emptiness where time and matter circle the drain into oblivion. The movie was once panned by the astrophysicist Neil deGrasse Tyson as being "the least scientifically accurate movie of all time" (Trzcinski 2020). Even so, "The Black Hole" proved highly successful at the box office; despite its cost (it was at the time the most expensive film Disney had ever produced), the film grossed more than $35 million as compared to a budget of about $26 million for production and advertising.

There's more than one way to profit off a vortex. Every year three million tourists make the trek to Sedona, Arizona, a small (population 10,000) desert town situated just south of the Grand Canyon in a mostly empty part of this mostly empty state. They come seeking not desert solitude but the town's mysterious, swirling centers of energy that are reputed to have restorative physical, even spiritual powers. (Locals insist that the proper plural of the Latinate "vortex" as it refers to their phenomena is not "vortices" but "vortexes.") Different vortexes advertised on the website "Sedona Red Rock Tours" are said to emanate "magnetic" (feminine) or "electrical" (masculine) energies, while others offer a combined and sexually balanced "electromagnetism." Most supposedly induce subtle but distinct physical sensations such as "buzzing," "tingling," or "heating." The electrical vortexes, for example, are said to provide a jolt akin to the spike in a double espresso. It is also said that these buzzes, tingles, and hot flashes somehow move those who experience them to a higher spiritual ground (sedonaredrocktours.com 2019).

The Sedona vortexes were popularized in 1980 by a local medium named Page Bryant, but many contemporary believers insist that the area had been known for centuries among Native Americans for its potent spiritual energies. Bryant herself received formal training in the vortexes' healing powers by an Ojibwe Indian called Sun Bear, but Sun Bear's custom of charging Bryant and other pilgrims $500 a pop for spiritual enlightenment has been criticized by the American Indian Movement (Chantae 2020). Before it was a modern-day Mecca, Sedona was *the* place Hollywood filmmakers went to film westerns. Among the classics of the genre shot here were *Riders of the Purple Sage* (1931), *Angel and the Badman* (1947, starring John Wayne), and James Stewart's *Broken Arrow* (1950).

The town acquired a new claim to fame when more than 5,000 latter-day Aquarians arrived in town to take part in a geo-spiritual event known as the Year of the Harmonic Convergence. (That would be 1987, for those of us who remain stuck in the Gregorian way of counting the years.) People gathered at the red sandstone rocks of Sedona (and numerous other sites worldwide) to celebrate a date which marked the final quarter-century of a long, 5,000-year cycle in the ancient Mayan calendar, and presaged, therefore, the dawning of a brand-new era to begin in 2012. The celebration was one of about fifty such festivals worldwide, conceived and organized by José Argüelles, an art historian living at the time in Boulder, Colorado. Argüelles' New Age writings and prophesies (*The Mayan Factor; Time and the Technosphere; Earth Ascending; Surfers of the Zuvaya*) brought together some strange bedfellows. The film actress Shirley MacLaine was on board with the movement, while Gary Trudeau satirized it in "Doonesbury" as "a moronic convergence," a "sort of a national fruit loops day." Newcomers to Sedona and numerous other international "power centers" identified by Argüelles sang, mingled, and prayed to the Aztec god Quetzalcoatl to stop the earth from spinning off into interstellar space as part of a new "galactic synchronization" brought about (in Argüelles' cosmology) by "random dissonant frequencies [caused by] radioactivity, chemicals, fluorocarbons."

Argüelles died in March 2011, a year before he might have assessed whether the new age he foresaw commencing in 2012 had come to pass. But because of all the publicity, the sleepy little town acquired a mystical *je ne sais quoi*, and Argüelles' prophesies firmly cemented Sedona's place among the world's most widely sought spiritual destinations, along with Machu Picchu, the Great Pyramids of Egypt, and Uluru (formerly Ayer's Rock) in Australia's Northern Territory (Niland 2012). Sedona's vortexes might be as worthless as the indulgences doled out by Chaucer's Pardoner, but that hasn't discouraged believers or stopped the town from peddling their wonders. The quasi-official website sedona.net, for example, promises that visitors to one of the nearby vortexes "may feel a range of sensations from a slight tingling on exposed skin, to a vibration emanating from the ground" (Chantae 2020). Especially common is said to be a tingling along the neck and shoulder blades. Dwight Garner, a writer for *The New York Times*, said he felt a "vibe in the air, something not quite audible, a kind of metaphysical dog whistle that calls people out to have a look around and to try to feel something that is hard to put into words" (Garner 2006).

"You feel like you don't belong there," said Elaine Lee, when I asked her about her impressions of the landscape. "It's like the pyramids, like Stonehenge, like dropping into a labyrinth. You think: 'My God, what is this place?'" When I talked with Lee, she had just completed a week-long, recuperative, post-pandemic tour of the American Southwest that included a two-day layover in Sedona where she had gone to assess the famous mise en scène. The experience left her unconvinced. "I felt nothing even remotely mystical," she said. "And the town doesn't do the landscape any favors. It's full of guru stuff, mostly silly."

What about the vortexes, the spiritual insights, the buzzing, tingling energies? Was there anything to all the publicity, the earnest testimonials? Lee laughed. "When I hiked to Bell Rock," she said, "I felt a tingling in my legs. But I felt the same thing walking around the town, so it was probably sweat evaporating in the super-dry air." "I also went to Boynton Canyon. I sat for an hour, trying to sense something. People all around me were saying 'I feel something.' But I really didn't. We kept getting interrupted by a man playing a flute, singing songs of love and healing, and passing out little rocks in the shape of hearts. When I asked him about the science of vortex healing, he immediately shut down. Still, it's a very interesting place. There's something about it, for sure. If you were part of a culture that valued coming-of-age rites or 'vision quests,' this would be where you'd go."

Tornadogenesis

At the top of the scale of earthbound vortexes are violent, cyclonic storms like tornadoes and tropical cyclones. (Tropical storms that develop over the Atlantic Ocean are popularly called hurricanes, although from a meteorological perspective they are tropical cyclones, no different from the typhoons that spawn in the northwestern Pacific or the cyclones common to the Indian Ocean.) The great destructive power of such gigantic storms is most often owing more to the flooding, tidal surges that accompany them as they make landfall than to outright wind damage, but now and then these oceanic tempests produce swirling winds almost equal in intensity to those inside the vortex of an EF-5 (Enhanced Fujita scale) tornado. On April 10, 1996, Tropical Cyclone Olivia topped out at 253 mph as it passed by Barrow Island off the northwest coast of Australia. Truly, there seems little to choose between Olivia's furious, sustained blast and wind gusts of 301 mph recorded on May 3, 1999, as a tornado passed over the town of Moore, Oklahoma.

Here's how the powerful, cyclonic storm known as a tornado forms. The circulation of air currents begins, always and everywhere, with inequities in the distribution of heat. Over sunny days in certain places—the open plains, a warm oceanic current, even a large asphalt parking lot—trillions and trillions of molecules of air gain heat energy, and as they try to come to some kind of thermal equilibrium in relation to their surroundings they start to move in large, energetic masses called "parcels." The more heat the molecules absorb, the more active they become, and as parcels of moving air shove and bump each other about, they expand, grow relatively lighter and more buoyant, and begin to drift upward. Tons and metric tons of this lively, heated air push their way higher and higher into the sky, and as the parcels rise, they take huge amounts of thermal energy along for the ride.

These warmer flows eventually collide with equally large masses of colder air miles high above them. At first the rising parcels have enough buoyancy to force their way upwards and through the cold air, but eventually they cool down, shrinking and condensing much of their vapor content into liquid water droplets. The individual droplets are small and light—20 microns, or about one-thousandth of an inch, would be a big one—but there are a staggeringly large number of them. Meanwhile, down near the ground, the steady, upward flow of warm air temporarily leaves behind an area of somewhat lower atmospheric pressure. Into this region is drawn even more warm air from the surrounding areas, further accelerating the upward-moving river of air and moisture.

Soon a boiling, churning cloud forms high in the sky where the parcels of warm and cold air meet; the stronger and swifter the updrafts, the taller and bigger becomes this "cauliflower." These immense, bulging clouds are commonly seen in summertime throughout much of the lower forty-eight states, especially in the American Midwest. Moisture turns them the color of gunmetal; at this stage the sky begins to look "unsettled" or even "angry," anthropomorphic terms that despite their impressionistic quality in fact seriously understate the explosive potential of the natural process that is playing out. The words merely hint at the inhuman amounts of energy coming into play as huge masses of air smash about in an attempt to reach some state of thermal equipoise.

Most often the atmospheric drama ends with sudden, heavy rain showers or thunderstorms. But now and then, to this gigantic cocktail shaker of warm and cold air is added a third ingredient, something meteorologists call "wind gradient" or "wind shear." The term describes what happens in the atmosphere when two adjacent layers or parcels of air happen to be moving at different speeds or in different directions, or sometimes both. "Shear" in physics is one of the different types of ways an object can be physically changed or "deformed." Anything that is pushed or pulled will tend to be correspondingly squished or stretched, even if only a tiny bit. When compressive or tensile forces are applied to opposite ends of the object, each side shifts relative to the other, giving rise to a tearing or "shearing" effect.

Set a book on a table and push laterally on the top cover. The interior pages will slide a little in the direction of force relative to the stationary bottom cover, slightly changing the book's shape. The book is now under "shear stress," and all objects will deform when this kind of lateral force is applied to them. (Even a brick will deform in this way, though to a much, much lesser degree than a book. The brick simply pushes back; it resists deformation until the shear force acting on it is strong enough to overcome inertia and slide the whole thing over the tabletop.) Unlike solid objects such as books or bricks, however, gases (such as air) are much more easily mashed and shoved about. When large masses of air are subjected to shearing forces, the result is more like pushing on the top of a loose stack of papers. Different parcels and layers of air readily squish, slide, and swirl past one another.

There are two types of wind shear—vertical and horizontal—and often they occur together. One of the places wind shear is commonly felt is near the outlets of canyons or around large buildings. If you have ever walked across an urban plaza toward some high-rise apartments and suddenly been tipped slightly off stride by an abrupt change in wind direction, or if, on a breezy fall afternoon, you heard sounds coming from around a corner, you have experienced some of the more common, down-to-earth effects of wind shear.

As you might expect, wind shear also develops frequently at the centers of thunderstorms when huge columns of rising and falling air alternately flow into or out of the same space. The sudden, often violent reversals of wind direction and velocity inside thunderstorms produce distinct "shear lines" that are particularly dangerous to aircraft that are powered down, flying slow, at altitudes near to the ground—as is always the case, naturally, when a plane makes its landing approach to a runway.

What makes a tornado out of this witches' brew of random, churning air currents? Now and then, shear winds inside the core of a developing thunderstorm happen to blow sideways alongside the huge columns of ascending and descending air, setting them slowly turning. It doesn't happen immediately, and thankfully it doesn't happen always. Getting an enormous, vertical column of air moving sideways is like trying to push a blob of Jell-O across a dinner plate. Mostly the mass just squishes and stretches. But sometimes, if the wind shear forces are sufficiently sustained and sufficiently large, eventually the entire column of air starts to spin around a vertical axis, the beginning of a true atmospheric vortex. This rotating cylinder of air is typically large—it can be ten miles or more in diameter—and at this early stage it's called a mesocyclone. These "middle" or "intermediate" cyclones are common—lots of times you can't even see or feel them—and they're typically part of convective storms and weather systems. The winds associated with mesocyclones at times can be gusty—about 40 miles per hour—but they're not particularly hazardous.

Sometimes, however, rotational kinetics take command of the situation. First, the vertical, slowly rolling tube of air starts to behave exactly like a top, a figure skater, or anything else that spins. As soon as the column of air begins to spin, in other words, it acquires angular momentum, and it wants, so to speak, to "conserve" whatever amount of that property it possesses. But that's not all that's happening. While the column of air begins churning slowly round and round, air pressure inside it continues to fall because the warm air inside it keeps rising skyward. Bit by bit, as the interior pressure drops, the once large, relatively slow-turning mesocyclone reconfigures itself into something known as a "dynamic pipe."

The amorphous, slow-turning washtub of air contracts severely to a more sharply defined tube about a mile across, and all the while that the rotating vertical column is collapsing inward, air inside it keeps being rapidly sucked higher. The spinning cylinder simultaneously stretches up and shrinks in, and the more it contracts in diameter (remember that spinning skater pulling in her arms?), the more its rotational velocity increases. This unseen drama often takes place high above the surface of the earth at the center of a towering, churning cloud called a super-cell. But if the tube of rotating air descends from the cloud that spawned it and contacts the surface of the earth—fortunately, most don't—it "touches down" and becomes officially a tornado.[1]

Atmospheric conditions and pre-conditions like these are com-mon on the plains of the American Midwest. More than a hundred thousand thunderstorms develop annually in the United States alone; so why, then, does only one in every hundred or so become a tornado? The answer depends almost entirely on chance, on the random motions of trillions upon trillions of air molecules whose individual behav-iors are as fickle as fame and fortune. If the world were a determin-ist sort of place, there would be little room for surprises: a cold wind streaming down from the plains of Alberta, some moist, oceanic air drifting north from the Gulf of Mexico, and hey! presto! there's your tornado. But not all swirling mixtures of cold and warmer winds grow up to become tornadoes. For that mercy we can thank Mother Nature's inbuilt unpredictability.

In the late 1700s, the French mathematician Pierre-Simon Laplace confidently envisioned a not-too-distant future when humans could predict future events with much the same accuracy as they solved a multi-step equation. All it would take, Laplace asserted, was assembling the right data. At the time, his confidence was not necessarily over-weening. Laplace was an expert in practical subjects ranging from oce-anic tides to the movements of planets (some called him "the French Newton"), and he was also a formidable theoretical astrophysicist, once suggesting that there might be stars so massive that not even light could break free from their gravitational pull. (In other words, Laplace imag-ined "black holes" more than a century before they were glimmers of possibility in Einstein's or other scientists' brains.)

But Laplace was, and is still, best known for his enthusiastic belief in scientific determinism. He proclaimed this grand vision first in his popular 1796 treatise, *The System of the World*, and returned to the subject three years later in the more expansive, five-volume *Celestial Mechanics*. As Laplace put it, the trick was to know "at a given instance of time, all forces acting in nature." For such a "being" (the term Laplace used was *une intelligence*), "nothing would be uncertain."[2]

Sentiments like these sound naive (if not ominously Faustian), but Laplace was not alone in his enthusiasm. All through the nineteenth century, western scholars and philosophers were in general agreement that it was only a matter of time before humans figured out everything there was to know about the way the world worked. Even well into the twentieth century, Albert Einstein, in the face of mounting evidence of quantum indeterminacy, insisted stubbornly that God, *Der Alte*, was not a gambler and would not play dice with His creation. Their optimism now seems to have been unfounded. Several generations of quantum physicists following the lead of Max Planck, Niels Bohr, Werner Heisenberg, and Erwin Schrödinger have told us it just isn't so. Some of what happens inside an atom is ineffable—not because the behavior of atoms hasn't been studied at great lengths, but because in the end, what goes on there depends in part on random chance.

As a teacher of English, I used to grump when television newscasters gave weather forecasts in terms of mathematical probabilities, as in: a 20 percent chance of snow, a 90 percent chance of rain. Such talk seemed to me a pretentious abuse of language. Why not just say that snow was possible, rain likely? Now I see the reason to give weather forecasts in terms of statistical probabilities. Deep down, nobody really knows what tomorrow will bring, and the best we can do is treat weather as quantum physicists treat electrons, in the aggregate rather than on any case-by-case basis. That negatively charged particle hovering around an atomic nucleus is almost guaranteed to be located where the electron cloud is densest; 99 times out of a hundred, there's where you'll find it. But you can't entirely discount the possibility that it might be somewhere else. Strange and improbable things do happen with weather as with atoms, even as once, on the morning of January 19, 1977, if you lived near the southern tip of Florida, you might have waked to this headline in a souvenir edition of the *Miami News*: *The day that couldn't happen: Snow in Miami!*

The Mysterious Pseudoforce of Gaspard-Gustave de Coriolis

During hurricane season, television weather forecasters routinely display the different storms with 6's and 9's to indicate their locations and tight, circling winds. The graphics accurately depict reality. Photographs taken from satellites directly above a cyclonic storm clearly show its pinwheel shape—a dense, circular core surrounded by tailing, spiraling wisps of clouds. Spreading over several hundred miles, the storms in such photographs look much like the Hubble Space Telescope's image of Messier 100, a magnificent spiral galaxy located 65 million light years distant in the in the Virgo cluster. Pictures of the galaxy and the cyclone set alongside one another show an eerie similarity, as if oceanic storms and galaxies, despite their almost incomprehensibly different scales of magnitude, had been shaped by the same artisanal hand.

With both hurricanes and galaxies, however, appearances are deceiving. The delicate, fleecy spirals belonging to stars and storms gives no hint of the huge amounts of energy packed into those swirling clouds. An average Atlantic tropical storm, for example, carries the equivalent of about 1.5 terawatts in rotating wind energy. Since a terawatt represents one trillion watts, a hurricane each day produces energy equal to about one-fourth of the world's overall electrical generating capacity (Donahue 2016). It's hard to imagine energy potentials as great as this. As for the energy potential packed into the spinning arms of Messier 100, to speculate about it is meaningless; it's impossible to express the "energy" of an entire galaxy in metrics that make sense.

So why do tropical cyclones invariably take on this lovely, spiraling configuration?[3] (We'll look at spinning galaxies in the next chapter.) The answer to a great extent depends on the influence of the whirling globe itself and a related phenomenon known as the Coriolis Force which steers the directional vector of masses of air sideways as they move across the surface of the earth. The Coriolis Force is well known to meteorologists for the significant role it plays in large scale weather phenomena, but it is often described in introductory physics textbooks as an "effect" or a "fictitious" force—if, indeed, such textbooks mention it at all.[4] Yet this "made up force" (in the words of one college physics textbook) appears wherever things happen to be spinning. Within any rotating reference frame—it can be anything from a children's merry-go-round on up to the Milky Way itself—the Coriolis Force causes objects to undergo lateral acceleration whenever they attempt to move in a straight path.

The Coriolis Force bears the name of Gaspard-Gustave de Coriolis, the person who first identified it in a mathematical paper of 1835 describing the transfer of energy in waterwheels. The son of a French sub-lieutenant who fought with General Rochambeau at Yorktown and the Battle of the Chesapeake during the American Revolution, and one of seventy-two scientists whose names are inscribed on the first floor of the Eiffel Tower, Coriolis was educated at the École Polytechnique in Paris, a school to which he subsequently returned as a young man to teach applied mechanics, mathematics, and physics. Coriolis seems all his life to have suffered from ill health. His students at École Polytechnique nicknamed him "death dodger," and no wonder: "Do not disturb me," he would intone when he greeted them each morning. "Who knows if I will be alive tonight? I feel my body faint, my mind is alone standing,

Aerial photograph of the swirling winds of Hurricane Isabel (2003). Superimposed on the image are vectors depicting the Coriolis force, which in the northern hemisphere accelerates moving air masses to the right of their forward velocity (Titoxd, Wikimedia Commons).

it may be annihilated sometimes with my last and weak forces" ("Biographies About the Eiffel Tower" 2020). His poor health may have deterred the young scholar from marriage, and it indeed took its toll soon; Coriolis died in 1843 at age 51, premature even for the times.

But whatever strength Coriolis had, while he had it, went to mind and work. His interests in the physics of motion were wide, to say the least; he wrote extensively on subjects ranging from waterwheels to the oscillation of pendulums, to the ebb and flow of coastal tides. Even when he took time off from teaching and writing to play a game of billiards with Monsieur de Tholozé, his supervisor at École Polytechnique, and François Mingaud, a one-time captain of infantry, Coriolis could not resist looking at things with an inquisitive eye. Captain Mingaud played billiards using a rounded leather cue tip to gain extra control over the spin of a ball when he struck it, and it was probably the captain's "extraordinary and surprising strokes" (Mingaud's words, taken from his autobiography) that inspired Coriolis to write what is still, nearly two centuries later, the definitive, mathematical analysis of pool balls bouncing around a table.

In *Théorie Mathématique des Effets du Jeu de Billard* (this book, like Coriolis's study of waterwheels, was also published in 1835), the frail mathematician took an exhaustive, clinical approach to a simple, recreational game, systematically laying out a series of equations that described the various and often strange paths taken by ivory balls rolling across a horizontal surface, rebounding off rails and off one another (Coriolis 2005). His failing health notwithstanding, Coriolis addressed and dispatched mathematically the multitudinous spins, collisions, and caroms of billiard balls so completely and methodically he might have been a sniper picking off a row of clay pots atop a fence.

Indeed, all things that went round and round seem to have held for Coriolis a particular fascination. During his tenure at the École Polytechnique, he created the first accurate mathematical principles for describing the velocities and accelerations of machines with spinning parts. Coriolis meant his equations to apply only to the rotary components of industrial equipment, but the rules he developed were subsequently found to apply equally well to all objects that are moving freely within any rotating (technically, any "non-inertial") frame of reference. It does not matter a whit whether these "non-inertial" spaces are very small or very large. A "Coriolis Force" acts on children as they clamber back and forth on a spinning schoolyard merry-go-round, and it takes effect on the huge rivers of air that flow constantly over Earth's oceans and land masses. Coriolis's equations even apply to the rippling swirls that form on the skirts worn by Sufi dancers and to the rotation of sunspots big enough to swallow a planet.

Here's a simple illustration. Say you're sitting next to the window of an Airbus 320 flying westward somewhere over Nebraska, looking down on a landscape covered by hundreds of those lovely green crop circles. (A great many Midwestern agricultural fields are made circular, of course, because the crops growing in them are watered by sprinklers spinning round slowly on center-pivot irrigation systems.) You glance down and spot a boy and a girl playing a game of catch on one of these circles. The boy stands close to (but not quite on) the center of the bright, verdant sward; the girl is positioned somewhere near its circumference. As you watch, the ball snaps back and forth between both players in a perfectly straight line. (From your position directly above the players, you cannot see its shallow, parabolic curve downward as gravity pulls the ball toward the earth.) Back and forth, back and forth, the ball flies as if traveling on a tautened string.

Now suppose that the entire circle somehow begins to rotate clockwise, turning round like a gigantic turntable. The players themselves are unaware the ground on which they stand has begun to spin, and they try to keep on with their sport. But they can't. A ball thrown outward by the boy now zips by his partner's right shoulder, well beyond her reach. From your point of view, looking down on the scene from high above in seat 29F, the reason for what just happened is obvious. As before, when the ball was thrown it still flew in a straight line across the circle. But because the girl is positioned close to the outside of the circle, you can see that she is standing much farther away from the circle's axis of rotation than is her partner. You can see, therefore, that she is moving much faster in a clockwise direction than he is, and you can see also that her sideways speed will be much greater than the sideways speed of the boy *and, therefore*, the sideways speed of the ball as it leaves his hand. So, it passes her by and disappears into a nearby cornfield.

But what about events as viewed by the players themselves? They understand their situation quite differently. They have no way of knowing that they are now standing on a spinning surface. From their point of view, nothing has changed; they still face each other in the same relative positions as before. Yet now, when they try to play catch, they both see something strange: as soon as the boy releases the ball, it accelerates mysteriously away from the girl along a *curving* path. Puzzled, they try again with a spare baseball. Same result: as soon as the boy throws the ball, it swerves sideways and vanishes. The players have no explanation for this weird phenomenon. From their point of view, the ball's tailing, lateral movement flatly contradicts Newton's first and second laws, which hold that bodies remain either at rest or in uniform, linear motion so long as no external, non-zero forces act on them. It's a hot, sunny afternoon just west of Omaha, with nary a breeze to ruffle the tassels on the corn. Yet clearly, *something* must be making the ball veer away from the girl like that. But what?

To preserve Newton's laws of motion in this and similar situations, physics texts commonly explain the ball's sideways trajectory as resulting from "fictitious" forces or "pseudoforces."[5] The terms are accurate but somewhat misleading; hearing them, many people go on to assume that such forces are wholly imaginary and that they belong, therefore, in the same fanciful category as Santa Claus and Vikings with horns on their battle helmets. In fact, the Coriolis Force is measurable and (so say some analysts, anyway) is an emphatically real force that appears invariably in all rotating frames of reference. The Coriolis Force always acts perpendicular to the axis of the rotating frame; the further away any object is from that axis, the greater the force that acts (or, *pace* Newton, *appears* to act) on it.

In other words, the Coriolis Force causes any object moving within any rotational frame to be accelerated sideways—that is, to be deflected from motion in a straight line—even when no external, non-zero force seems to be causing it to do so. In still other words, Gustave-Gaspard Coriolis proved that if Newton's equations were to be accurate in describing the motions of objects situated within any rotating frames of reference, a *new* inertial force, proportional to the rotational (angular) velocity, acting sideways and in the direction of rotation, had to be included in any calculations.

Since we live in a rapidly spinning (i.e., "non-inertial") reference frame, you might wonder why you were made to learn Newton's first law of motion; strictly speaking, it's not valid on the Earth (or anywhere else in the observable, wildly spinning universe, for that matter). The answer is that the effect of the Earth's rotational speed on daily activities is usually so trivial that it can be safely ignored. Most of the time, we plan visits to friends and shopping trips without taking into our calculations that we move hither and yon across the surface of a rapidly spinning planet in a fantastically spinning cosmos. But now and then we see face to face.

I remember the first time I witnessed a total solar eclipse. It was on the morning of February 26, 1979, and my wife and two young daughters and I stood atop a small hill near the municipal water tower on the outskirts of Pullman, Washington, where I was attending graduate school. We had made all the usual preparations for watching an eclipse, and we were excited to think about peering through smoky goggles to see rare celestial phenomena—the sun's pulsing corona, Baily's beads, distant stars shining brightly at mid-day. But nothing at all had prepared me for the best part of the show which, I was to discover, was not *up there* but *down here*.

A few scant moments before the sun's disk went black, from out of the west there appeared without prelude a deep and terrifying blackness. It was, as I came to understand later, the approaching umbra cast by the moon careening in its orbit around the earth, even as the earth, in turn, was spinning beneath it. The shadow was several miles away when I first saw it rise above the horizon. Coming straight at us, it moved too fast for action or thought; I simply froze. All around me I heard screams; indeed, no other response seemed appropriate. I felt as if I was living my last seconds on earth. Waiting for the celestial show earlier that morning, I had what I thought was a healthy skepticism regarding Angels of the Presence. But during those scant few seconds before that darkness rolled over me, I understood what it meant to be sore afraid. There is simply no way for the mind to process a wall of shadow, wide as the world and high as the sky, spinning toward you at 2,900 miles an hour.

If the earth were somewhat more sluggish in its diurnal rounds (as is the case on Venus, for example, where a single "day" lasts from January to September), atmospheric winds would blow straight from the poles toward the equator, and large, cyclonic storms that developed over the tropics would be little different from those that form anywhere else. Hurricanes and cyclones would take shape much like willy-willys, tornadoes, and waters disappearing down a toilet bowl, spinning sometimes in one direction, sometimes in another.[6] But close to the favorite breeding ground for hurricanes, near the Cape Verde islands west of the continent of Africa, the earth is rotating eastward at close to a thousand miles an hour, and so a Coriolis force is constantly accelerating sideways any large air masses blowing across its surface from north to south, curving them in the beginnings of a huge, counterclockwise rotation. (To say these winds are being accelerated does not mean that they are increasing in velocity, only that they are changing their directional vector, which properly counts as an acceleration.)

The situation is the opposite below the equator, where winds blowing mainly from the south acquire a slow, clockwise spin. This is why photographs of Tropical Cyclone Olivia (1996) taken from satellites above the Indian Ocean show a tight, right-hand spiral of clouds, while pictures of Hurricane Dorian (2019) parked over the Bahamas showed similar bands of clouds swirling in retrograde motion.

A Coriolis acceleration also changes the directional vector of falling objects—though to a much lesser extent. The higher you venture above any single place on the surface of Earth, the greater your tangential velocity with respect to its axis of rotation, and the greater, therefore, the potential Coriolis acceleration. Climb a beanstalk all the way up to one of NASA's geosynchronous satellites and gently let go of your cellphone. It will hover right beside you, zipping eastward at 7,000 miles per hour. But if you were to try to hurl the device straight down, a Coriolis acceleration would cause it to land miles to the east of the beanstalk's roots.

Even the tops of tall buildings spin around earth's axis of rotation with ever-so-slightly more speed than their bottoms; the top and bottom of a skyscraper have the same angular velocities but different tangential speeds. This slight difference can be responsible for amusing (if negligible) consequences. Here's an interesting example. On August 21, 1908, Charles Evard ("Jabberin' Gabby") Street, the catcher for the Washington Senators (since 1961, the team has been known as the Minnesota Twins), agreed as part of a publicity stunt to catch a ball dropped from the top of the Washington Monument.[7] The first ten attempts miscarried because the balls kept bouncing off the lower, wider base of the monument. At last, someone on the observation deck hurled a baseball far enough outward to clear the bottom granite blocks. Gabby, eyes straining upward, buffeted by strong wind gusts, couldn't spot the ball at first; he caught sight of it only after it had fallen more than halfway down.

A second or two later, it smashed into his mitt with an impact force calculated at 300 pounds and a sound, according to one eyewitness, "like a rifle shot" (Sharp 2021). "I felt it right down to my heels when it hit!" babbled the giddy, triumphant catcher. It was a heroic day for the game of baseball, and—so far, anyway—a stunt never again to be attempted. Amid all the excitement, however, Jabberin' Gabby probably didn't notice that during the 5.9 seconds it took the ball to fall to earth, he had to shift his mitt a few inches further east to keep it centered on the rapidly descending spheroid.

Cyclones and baseball stunts notwithstanding, the force first described by Gustave-Gaspard de Coriolis has not been uniformly disruptive to human affairs. Far from it. For many centuries, Coriolis accelerations were the main drivers of economic and demographic "globalization," reliably deflecting strong atmospheric winds which otherwise would blow over the oceans in predominantly northerly or southerly directions. In other words, this so-called pseudoforce is a real-world engine that for hundreds of years made possible—for better, and, lamentably, sometimes, for the worse—the eastward and westward transit of sailing ships between ports scattered throughout Europe, Africa, Asia, and the Americas.

Without Coriolis accelerations, there would be no prevailing winds blowing across the earth in latitudes extending from Egypt, Florida, and the East China Sea in the north, or, below the equator, to Brazil, Australia, and the Republic of South Africa. These rivers of air, of course, are known as the "trade winds," and they have been used by mariners ever since late medieval times when Portuguese sailing ships began to rely on such them to navigate back and forth between the coast of Africa and European ports. (The name "trade winds" combines two loosely related etymologies: it comes first from an early Germanic root meaning "track" or "to tread a path," but as the winds' role in international commerce became increasingly important, "trade" winds took on mainly economic significance.)

Even now, oceangoing ships sometimes take advantage of those steady, subtropical winds to save fuel or to hasten their way toward ports. But vessels that rely for movement solely on the fantastically useful Coriolis force exist now only in paintings, or for sport, or for training annually a handful of Coast Guard recruits. Over the last half of the nineteenth century, hundreds and hundreds of graceful, tall ships were gradually phased out of commercial maritime service, replaced by ships driven by massive underwater propellers. One of the last true mercantile sailing ships, the beautiful, four-masted barque *Pamir*, carried her final cargo round Cape Horn in January of 1949. It's surprising she lasted so long; even so, it's important to note that the screw-steamers that replaced her, far from being a new technology, basically exchanged one source of rotational energy for another, more obvious one.

Trade winds and ships' screws are reminders of how much useful energy can be extracted from things that spin. The kinetics of both forms of energy production are the same, only their hierarchy is different. When, in 1970, my wife and I sailed across the Pacific Ocean from Sydney to San Francisco, the SS *Arcadia* was powered by engines whose rotary lineage could be dated back to the first-century mathematician Hero of Alexandria and his toy *aeolipile*, a bladeless, spinning turbine which worked by means of steam jets blasting tangentially from nozzles on the circumference of a container full of boiling water.

Little Twisters

Now and then you can spot small, lonely whorls of dust skittering across an open field, or maybe you have seen bits of paper whirling together down a city street on a hot summer afternoon. These dust devils, as they are commonly known, appear almost as if by magic. Dwarf tornadoes are almost always harmless curiosities (though now and then one comes along that is big enough to do some serious harm to life and property). Yet they occur so randomly and so suddenly it's easy to be spooked by their playful, silent magic, and for that reason in almost every culture they carry occult meanings. Americans most often call them dust devils, as if the mysterious little vortexes were up to some unspecified mischief. To the early Navajo, on the other hand, they were *chindi* ("ghosts"), omens either of good or bad fortune for the observer depending wholly on which direction they happened to be spinning when first spotted.

Contemporary Australians use the term *willy-willy*; the term sounds frivolous, a word one might use to refer to a thing of no account, but that popular tautonym carries with it ancient aboriginal connotations of a mischievous spirit who hides in the twisting column of dust and whisks away naughty children. In France, writes Lyall Watson, these whirling, tornado-like shapes are *trombes giratoires*, while in the deserts of the American southwest their graceful, swaying vortices are known as sand augers (Watson 1984, 62). In the east African countries where Swahili is spoken, a dust devil is known as a *shetani*, a word that describes a shape-shifting demon or spirit, especially a mysterious, intransigent adversary who can never be overcome, only avoided.

In the deserts of Ethiopia, writes the Swiss traveler John Burckhardt, these "prodigious pillars of sand … [stalk] with majestic slowness … their tops reaching to the very clouds" (Brown 1962). And throughout the Middle East, where extremes of heat and convection sometimes create writhing columns of sand and dust that resemble sinuous, living creatures, the spinning clouds are called *djinn* to identify them as one of the three orders of living beings. Along with humans, whom He made of earth, and angels, who were made of light, Allah created *djinn* from smokeless fire.

Vortexes ripple and swirl around us daily, but normally they're so small we don't notice them. Yet one of the most common of these little twisters is potentially dangerous. It's called a Kármán vortex street (from Latin *stratus*, pple. of *sternere*, to spread, lay down, or pave), and it describes a repeating series of swirling currents that form when winds blow downstream from blunt objects. A good place to spot these graceful, circling eddies is in the waters trailing away from bridge abutments. Speeding motor vehicles also whip up these powerful little vortexes—an Interstate highway is a veritable cauldron of them—but unless you happen to need to change a tire by a busy roadside, you can't feel them. In the late fall, though, sometimes you can spot Kármán's vortexes magically uplifting swirls of leaves as vehicles hurry along tree-lined country roads.

Kármán vortex streets also reveal themselves in satellite images of cloud systems that form over relatively isolated land masses such as islands or lone mountain peaks, and they sometimes occur in places and situations where they can cause several undesirable effects. Named for their discoverer, Theodore von Kármán, a Hungarian mathematician and research scientist, the phenomenon covers a variety of curious sounds and flutters that come from all kinds of things ranging from submarine periscopes to skyscrapers. (It's a small world, fluid dynamics: Kármán was one of Ludwig Prandtl's students in his laboratory at Göttingen.) These little twisters even form when wind blows past car antennas and telephone wires. It was likely vortex street vibrations, for example, that were responsible for the sounds Glenn Campbell heard singing in the overhead wires in his plaintive 1960s ballad about a Wichita lineman.

On a much larger scale, the same swirling currents have been known to cause serious instabilities in tall structures such as radio towers or industrial chimneys, carrying sufficient energy potential even to down them. This happened in 1965 to two of three cooling towers at the nuclear electrical generating facility at Ferrybridge, Yorkshire. Investigations into the collapse of those towers determined that design engineering stress calculations had been based only on wind effects on a single tower, ignoring the much greater forces that could buffet any structures that had the misfortune to be located downwind, smack in the center of a swirling Kármán vortex street.

At other times, nature's spiraling air currents are benign. Every two years, a couple of dozen model aircraft builders gather on a large meadow somewhere in Europe to compete in the F1A international towline glider championship. Like their full-size counterparts, these slim, elegant sailplanes lack engines; to fly, they must be towed into the air like kites. Full-size gliders are most often hauled aloft behind repurposed Cessnas and a rugged, one-ton test nylon rope, but the small, featherweight replicas in F1A competitions require only 50 yards of Kevlar thread and a healthy pair of legs. Once the little craft are high overhead, modelers lead them around the sky like handlers at a dog show, searching for a thermal. Thermals are large pockets of warm, rising air. They form either as continuous, spiraling chimneys or as a series of spinning, ring-shaped bubbles that break off at intervals from the ground and drift upward. If we could see them, they would look like giant, rotating smoke rings. But modelers can't really see them, of course, so they depend instead on their aircraft to feel the bubbles for them. Patience and a light touch are the keys to success; it's a little like waiting for a fish to nibble a lure. The sensation is subtle, and when it comes it may be nothing more than a gentle bump or a persistent, barely perceptible tugging on the wrist. But when you first feel your creation being picked up by some mysterious, invisible hand, it's really a kick. You are about to soar like an eagle, if only empathically.

High above the Palouse grain fields of eastern Washington late one August morning in a brand-new century, I felt firsthand that same thrill as the Cessna 182 I was flying in bumped and lurched in and out of numerous, powerful atmospheric updrafts. Dry landscapes like the one we were flying over that day are especially suited to the production of large, booming thermals, formed when rising ground temperatures begin to heat the air directly above them. It's a simple, straightforward thermodynamic event.

As with dust devils, the process of forming thermals begins with untold billions of air molecules being energized by heat from the sun. The sun's radiation warms the surface of the earth, but—and this is critical—not at all evenly. Darker areas like plowed fields or asphalt highways absorb more heat more quickly than, say, bodies of water or croplands, and the heat absorbed by the darker areas on the ground will be transferred to the air molecules immediately above it.

Like bubbles in a pan of water near the boiling point, the heated molecules begin in ever greater numbers to expand. These warmer parcels of air become less dense, and, as the differential heating continues, they coalesce and begin to drift upward in large, towering plumes. Like dust devils, thermals usually contain significant rotating elements. But whereas dust devils are typically evanescent phenomena, the size and duration of thermals can often be much greater—so much greater, in fact, that the potent, twisting energies contained in them are large and persistent enough so that over time both humans and animals have learned to make use of them.

Birds long ago mastered the trick of using the wind's energy to conserve their own. Off to the west that morning and slightly below the aircraft I could see a few turkey vultures patrolling for carrion or (more likely) roadkill, wheeling lazily on extended wings. The wings of *Cathartes aura* are among nature's most efficient aerodynamic designs, giving up a miserly one foot in altitude for every twenty traveled forward, but I was surprised to learn that man-made wings are at least as stingy. Even a lumbering behemoth like Boeing's 747, for comparison, glides forward seventeen feet for each foot it drops, and many commercial gliders—contemporary planes like those built by companies such as Schweizer, Grob, and Blanik, for example—more than double that ratio.[8]

As soon as a bird or an airplane starts to move forward, air flows simultaneously over and under its wings which immediately begin to generate lift by creating differential airstreams on its upper and lower surfaces. At the same time a vortex is constantly forming just behind the tips the wings, spinning off in the opposite direction to the flow of air. This spiraling airflow (one forms behind each wing) is called a tip vortex, a kind of miniature, horizontal willy-willy. Tailing behind the wingtips in long, backward-streaming funnels, these graceful, swirling rings are commonly visible from the windows of almost any large aircraft that takes off or lands in damp or misty weather. I once sat next to a first-time flier who was visibly alarmed when we started our takeoff roll to see long, smoke-gray rings streaming off the plane's wingtips. She supposed the aircraft had suddenly caught fire. A kind and knowledgeable flight attendant explained to her that wingtip vortices were both natural and harmless. They were, she told her skeptical passenger, visible proof that the wing was doing exactly what it was supposed to do.

These tip vortices are usually invisible. But because they persist for great distances behind aircraft, especially behind large jets, they can be as hazardous as a malevolent *djinn* to smaller, lighter planes which can be flung violently out of equilibrium if they accidentally fly into them. When the Airbus 380, the world's largest passenger-carrying airplane, was in its design phase, engineers were so wary of the vortices that would spin off the behemoth that they stuck little V-shaped fences on the wingtips to help corral the dangerous, swirling currents.

Pilots try to avoid flying into these vortices. Remarkably, however, some species of birds have figured out ways to put these spiraling, turbulent airstreams to practical use. "One reason that geese and pelicans fly in neat V-formations," says David Alexander, "is that by maintaining a very precise spacing, birds can get a slight upward boost from the trailing vortex of the bird in front" (Alexander 2002, 29). Because the lead flyer in a V-flock gets no such benefit, however, the birds are apparently smart enough—and, remarkably, sufficiently fair-minded—to take turns up front to give the foremost bird a well-deserved rest. Exactly how much these bird-brains know about tip vortices is not clear; one might also ponder what motivates them to use that knowledge to act with something that looks very much like altruism. The birds' behavior suggests, in the words of primatologist and ethologist Frans de Waal, that humans are not smart enough to know just how smart animals really are (de Waal 2017). Alexander, for one, is convinced that an intuitive grasp of aerodynamic physics is the reason that birds flying in V-formation have been often observed exchanging the lead position.

Natural-born Helicopters

Birds and Blaniks aren't alone in making use of spinning air currents; even some members of the kingdom Plantae get in on the benefits. One of the cleverest are the different kinds of plants and trees that produce seeds equipped with tiny "wings." These samaras (as the distinctive, winged seeds are properly called) have evolved to spin as they fall to the ground. The wing-like structures produce aerodynamic lift when they move through the air; the resulting buoyancy slows and skews the descent of the seed as it descends, allowing for wider and more random dispersal.

Some species produce samaras of amazing aerodynamic functionality; the seed of *Alsomitra macrocarpa* (or Javanese cucumber), according to Alexander, when it is airborne "uses some of the same stabilizing techniques that human engineers use to stabilize flying-wing aircraft ... [including] swept-back wings with dihedral and reflexed trailing edges" (Alexander 2002, 49). Most samaras are much cruder in design and simply spin (or autogyrate) rapidly when they drop. Much beloved by children lucky enough to have access to one or more *Acer* trees in their back yard or on the school grounds, the samaras of the maple tree initially form as pairs of seeds growing from a single stem; attached to each seed is a paper-thin, teardrop-shaped "wing" that droops down and away on either side. In the spring (or sometimes in late summer or fall), the stems holding the samaras detach themselves from their branches, launching the tiny wings into the air, singly or in pairs. The samaras' graceful, twirling movement as they fall to earth, in combination with their distinctive, winged appearance, makes them appear, under breezy skies, to be swarms of large insects, perhaps locusts or grasshoppers, buzzing about. From a distance and with a little imagination, the seeds really seem to behave as if they had somehow learned the secret of animation. So little do the maple tree samaras behave like proper plants, in fact, that they are known popularly by un-plant-like nicknames. Attached to the tree or lying on the ground, they resemble "wing nuts," hardware fasteners with two small ears attached to them to facilitate hand tightening, and when airborne, these little "helicopters" might have been an early design sketch for a Sikorsky or Bell whirlybird.

As samaras hang on the tree to dry, in some species a fissure opens until the paired wings split apart and fall to earth as single seeds attached to the ends of single wings. This asymmetrical, "dumbbell" shape causes the samaras to fall in a spiraling, downward movement. The samara's center of rotation aligns with its heavy, seeded end, and so when it falls it spins in an ever-tightening helix, gaining rotational velocity in flight because its initial angular momentum is conserved. This spinning descent, it turns out, is slower and a good deal more efficient than mere free-fall. The descent of these drunken, spinning samaras has been clocked at less than half the rate of non-turning seeds.

Other species of trees produce tiny, unpowered helicopters of great complexity and mystery. The hornbeam tree, for example, goes the maple tree one step further. Unlike maple samaras, whose wings are flat, the hornbeam turns out samaras with true airfoils whose surfaces are cambered, convex on the upper side and concave on the lower. No matter how often one flings a handful of hornbeam seeds upward, they will, like falling cats, swiftly turn themselves right (convex) side up as they autogyrate toward the ground. Some species of pine trees produce also winged seeds that are amazingly efficient. They have asymmetric designs, with a single wing growing off just one side of the seed, so that as soon as it starts to drop it begins to spin rapidly. As the samara falls, spinning, through the air, a tiny vortex forms just above its leading edge. The more rapidly it spins, the greater the vortex that develops and the greater, consequently, the amount and velocity of air that is drawn in to fill it. This rush of air molecules along the upper surface of the seed/wing in effect "lifts" or "sucks" the samara upward, delaying its descent.

It's a remarkable example of aerodynamic efficiency and the trick of getting something for almost nothing. Spinning creates sufficient buoyancy to ensure that most of the seeds will be buoyed on the breeze far enough away from the parent tree to escape its umbra of growth-killing shade. Even the slightest wind currents set them drifting; on a late May afternoon in downtown Moscow, Idaho, I have seen these tiny, single-bladed 'copters piled along streets and sidewalks many yards distant from the tree from which they had earlier been dropped. "They take to the air," writes Watson, "like swarms of butterflies, and have even been known to fall on the decks of ships several kilometers out at sea" (Watson 1984, 172).

Samaras attract all sorts of admirers; one can purchase packets of samaras to amuse family and friends who might be so unfortunate (in the words of an advertisement I found online at buyflorals.com) never to have had "the pleasure of lying under a gigantic maple on a breezy autumn day." Buyers are encouraged to fling the samaras skyward in their own backyards, to drop them by the dozens from staircases and balconies at parties, or to toss handfuls of them at the bride and groom at the conclusion of their wedding ceremonies. These many thousands of falling, spinning wing-nuts would (or so the advertising copy reads) commemorate the occasion with a more elegant aerial display than a few flung handfuls of rice. As an added benefit, proponents claim, the celebratory samaras are less likely to dehydrate birds that might ingest them. The aerodynamic efficiency of maple tree seeds hasn't been lost on rocket scientists. When, at some time in the clearly foreseeable future, the first manned probes drift gently downward toward the surface of Mars, the vehicles' design just might make use of the simple, elegant autorotation technique modeled after those spinning samaras (Live Science Staff 2009).

CHAPTER 9

Really, Really Big and Really, Really Small

Heavens Above, Earth Below

> A slumber did my spirit seal;
> I had no human fears:
> She seemed a thing that could not feel
> The touch of human years.
>
> No motion has she now, no force;
> She neither hears nor sees;
> Roll'd round in earth's diurnal course,
> With rocks, and stones, and trees.

By the time the poet William Wordsworth composed the five short lyrics known collectively as his "Lucy Poems," the Earth's place and movements in the heavens had ceased to be a subject for debate. That the world was wide *and* round had been generally assumed beginning with the ancient Greeks. For Pythagoras (c. 500 BCE), a sphere was the simply most perfect of all shapes; no other shape for the world would do. In *Timaeus* (c. 360 BCE), Plato states: "the Creator made the world in the form of a globe, round as from a lathe." Plato must have imagined a spinning lathe (τόρνος) as the best way for a Creator to achieve regularity in form, or what we would call "point symmetry." A century later, by comparing the length of shadows cast by the noonday sun at two different locations a known distance apart, Eratosthenes of Cyrene had calculated the Earth's circumference to a degree of accuracy of about ten percent. There was simply no reason to believe the world was anything but spherical. Writes the historian Jeffrey Burton Russell: "No educated person in the history of Western Civilization from the third century B.C. onward believed that the Earth was flat" (Blakemore 2018). Nor was heliocentrism a new idea when Copernicus published *De revolutionibus orbium coelestium* (*On the Revolutions of the Heavenly Spheres*) in 1543.

Similar, sun-centered theories of the cosmos had been proposed repeatedly by philosophers and mathematicians from antiquity throughout the Middle Ages.

That a spherical world spun round on a central axis, however, was another matter entirely. To picture a spinning earth somehow seemed contrary to sense and sensibilities. Ptolemy, for example, had argued in *The Almagest* that if the earth's rotation caused the succession of day and night, it would be moving so fast that objects in the air (birds and clouds, for example) would always seem as if they were being left behind. (Ptolemy's descriptions of the cosmos were spectacularly wrong, of course, but he did much better with his accounts of music and optics.) The earth spins around amazingly fast, all right; at the equator, west chases east at close to a thousand miles an hour. But the atmosphere and everything in it simply rolls merrily along with rocks, stones, and trees.

Let's not be smug, however, in dismissing Ptolemy for not perceiving that the world turns. Figuring out Earth's spherical shape is comparatively easy because the evidence is abundant. At any seaside, for example, or on the shore of any large body of water: cast your gaze out to the horizon. The surface of the water looks wide and flat, and on it the boats move to and fro. But study the picture for a while and you will likely notice something peculiar: on the distant horizon, ships' masts and bridges can be seen clearly, but only as the vessels draw nearer to shore do their decks, gunwales, and hulls come into view.

You can observe the same thing if you move across a plain toward distant mountains. Bare peaks and rocky upper slopes are clearly visible from far off, but to see the trees and dense foliage crowding the base of the range you must be much closer. The only possible explanation for this phenomenon can be that the surface on which things move is *not* flat as it might first appear but slopes gently downward and away. A surface that is forever curving downward like that must sooner or later bend back on itself to complete a circle. Eureka! The world is *not* flat, but round!

Detecting that the world is *spinning* is another matter entirely. A day lasts the same amount of time whether you're located in Bogotá or Barrow. Stand still for twenty-four hours in either place and you'll circle back to where you started in relation to the sun. But how fast you make that turning from east to west depends entirely on how far you are situated from the Earth's axis of rotation. Persons on or near the equator are being whirled round the fastest; strolling the streets of Quito or Nairobi people are being spun easterly at velocities approaching that of speeding bullets. Sitting inside Santa's workshop, on the other hand, you spin at a snail's pace. Literally so: even a garden snail slouching along the ice at an average speed of three-hundredths of a mile per hour could run circles around people sitting in the Amundsen-Scott Station in Antarctica, located a scant two-tenths of a mile distant from earth's rotational axis, crawling along at a tangential speed of about 0.00005 mph, almost six hundred times slower.

Late morning, January 30, 2019. I watch a band of sunlight creep slowly down the frost that covered the windshield on my pickup truck. At 10:32 a low winter sun first kisses the top of the glass at the roof line. Minute by minute, the sun lifts slowly in the sky; minute by minute, more and more glass lies bare as the sharp, bright line of sunlight advances downward. I stare hard at the sparkling line where light meets frost; I see the glare move slowly downward, precise as a surgeon's blade. Things are in motion, no doubt. I know full well that it is not the sun that is moving. It's me, the truck, the globe that's spinning. Yet my brain is utterly unwilling to set aside what my eyes are telling me. I cannot believe otherwise: the sun is climbing up the sky.

The spin of the Earth divides day from night, separates the soft, hesitant luminosity of twilight and dawn from the brilliant glare of high noon. Just as important as God's dividing (in the biblical account) of light from darkness, however, are several less well-known effects of planetary spin. Many breaks had to fall our way for life to appear and to flourish on the Earth, and among several crucial events in the history of the planet was the forming of different layers inside it. The world is built much like a baseball; it has a distinct core, multiple surrounding laminates, and last of all, a cover. At the center of a baseball is a small, hard, rubberized cork nugget, and deep inside the Earth is a correspondingly diminutive, dense, and amazingly hot interior sphere. If you sliced Earth in two you would see this little ball within a bigger ball, made almost entirely of crystallized iron, looking just like the "pill" at the center of a baseball.[1]

This "inner core" is only about one-tenth the overall diameter of the planet, but temperatures there are fiercer than the outmost layers of the sun (about 10,000 degrees). The heat in the core is more than sufficient to liquefy just about anything, but pressures there are so extreme that the metal instead has properties more like a solid or plasma. This inner core rotates eastward along with the rest of the Earth, but—this has been discovered only recently—for unknown reasons it spins ever-so-slightly faster than the outer layers of the planet, completing an additional rotation about once every millennium. This dense, iron sphere slips and slides inside a gloppy layer of nickel-iron, a molten, "outer core" about 1,500 miles thick. This viscous region is in a constant state of turbulent convection; somewhat like the albumin within a raw, spinning egg, the outer core sloshes around and about the inner core, spinning at a slightly different rate. Now and then, for unknown reasons, it commences spinning in an opposite, westerly direction.

Both inner and outer cores, being mostly iron, are being constantly pushed and pulled by Earth's overall magnetic field; this magnetic field, in turn, is a byproduct of electrical energy that is generated continuously by currents moving through the liquid nickel-iron. Like the convection swirls rising in a bubbling pot of oatmeal, these buoyant, spiraling currents are perpetually in motion. The heat that drives them is thought to originate both from slow, thermal cooling of the inner core as well as from latent heat given off as the liquid nickel-iron shell gradually (as in millions and millions of years) solidifies.

The molten currents drift outward and upward toward the surface of the Earth, much like thermals above a wheat field on a hot summer morning, and, as they rise, the spinning of the planet (in conjunction with ever-present Coriolis forces) causes them to be helical (Glatzmaier 2020). It is these *interior* spinning currents of molten metal, suggests Gary Glatzmaier of the University of California at Santa Cruz, that initially created—and have subsequently maintained—our home planet's geomagnetic field. It's possible they're the reason we're here in the first place. Were the Earth's magnetic field not being continually recharged from deep within by these spiraling, upward currents, says Glatzmaier, its protective shield would have faded away a mere 20,000 years after the planet's formation, which is way too short a time to allow for the emergence of prokaryotes, T-Rexes, saber-toothed tigers, and you and me.

The bountiful combination of spinning, churning electrical and magnetic fields and forces in effect makes the planet one humongous electric motor. From an engineering standpoint, the planet can be called a vast "geodynamo" whose powerful electromagnetic field surrounds it and—o brave new world, that hath such good fortune!—deflects otherwise lethal solar radiation. Not so lucky was the planet Mars, which like Earth is positioned in a stellar "Goldilocks Zone" where ambient temperatures are suitable for the development of life. Scientists now believe that eons ago Mars lost much of its own protective magnetic shield when for some unknown reason or reasons its internal dynamo failed. The planet Venus, too, is theoretically in the biologically habitable zone, but its scorching surface temperatures and hostile atmosphere may be related to the planet's fatally sluggish rotational velocity (about four miles an hour, or one full turn every 243 Earth days). Without spin, Venus is simply incapable of generating and maintaining its own magnetic field (Choi, 2020).

Eudoxus (*On Speeds*, third century BCE) regarded the planets, sun, and stars as positioned on a set of twenty-seven concentric, nesting spheres, being eternally whirled in circles round a spherical Earth. Wherever the center of the heavens' rotation might be, we know now, it's not us. Yet apart from that single misconception—admittedly, it's a whopper—Eudoxus wasn't wrong. Every two minutes the earth has moved 2500 miles as it orbits the sun, while the sun has moved 20,000 miles as it spins around the distant center of our galaxy. In a 70-year human lifespan, the sun moves 300,000,000,000 miles. Yet this vast path is only a tiny arc of the circuit round the galaxy it takes the sun 200,000,000 years to complete. It's all moving, revolving, spinning; "within a galaxy," wrote the astronomer Vera Rubin, "everything moves" (Panek 2011, 36).

Rubin's data proved what George Gamow had once merely speculated in the journal *Nature.* In 1946, in a brief but provocative query letter to the editors of *Nature*, Gamow wrote that "one of the most mysterious results of the astronomical studies of the universe lies in the fact that all successive degrees of accumulation of matter, such as planets, stars and galaxies, are found in the state of more or less rapid axial rotation.... The rotation of stars themselves ... can be presumably reduced to their origin from the rotating gas-masses which form the spiral arms of various galaxies. But what is the origin of galactic rotation?" The answer, Gamow mused, might be that the universe itself was in a state of rotation; around what, we didn't (maybe could not ever) know. "What if," he speculated, "we took the way solar systems rotate and applied it to how galaxies move in the universe" (Gamow 1946)?

Early astronomers may have been wrong in their descriptions of the universe as a collection of nine, or twenty-seven, or fifty-six concentric, revolving "spheres." It's simply the turning of the earth that gives the illusion of stars, planets, and sun circling across the sky. Those first stargazers bollixed up celestial distances and relationships so completely that it's easy to overlook what they got intuitively, completely, spectacularly right: the essence of the cosmos is *rotation*. Even the Milky Way—which until about a hundred years ago was believed to comprise all the universe there was—it, too, turns round. Immanuel Kant had guessed it: in 1755 in his *Natural History and Theory of the Heavens*, Kant supposed that the Milky Way had coalesced from a diffuse cloud of gas, the pull of gravity collapsing, compressing, setting it spinning. Forty years later, in his five volume *Méchanique Celeste*, Pierre-Simon Laplace did the math. It all was wheeling, turning, every bit of it, from galactic clusters all the way down. Power, beauty, the sheer, amazing size, and exquisite choreography of it—how could you lift your eyes up and not marvel?

Not all celestial bodies spin (though most, in fact, do). But what got them started spinning in the first place? There's a clue to the answer in the way that we know most stars and planetary systems were formed out of huge collections of gas molecules, mostly hydrogen and helium atoms along with a much smaller (but still gigantic) proportion of elements like oxygen, carbon, sodium, iron, and all the rest of the 94 naturally occurring elements, all of which were created by a process known as stellar nucleosynthesis. The world, us, and everything else in it, as the astronomer Carl Sagan liked to say, are made of star stuff. Much of cosmologists' picture of the formation of galaxies and galactic clusters is speculation, of course; even the outlines are dim. But the physical similarities between the shapes of massive galaxies and lingering eddies in streams are suggestive; they lead one to ask, writes P.J.E. Peebles of Princeton, "whether galaxies might be the remnants of turbulence in the early universe" (Peebles 1993, 541).

That most galaxies rotate around a center is likely a natural consequence of the way they were formed in the first place. A galaxy begins as immense, inchoate, drifting clouds of hydrogen gas molecules. When such wandering gas clouds start to congregate, they're basically shapeless, aimless blobs. Blobs huge beyond all imagining, granted, but still something that behaves according to unvarying laws of gravity and rotational kinetics. Primeval "subclumps" of matter, in this scenario, accrete slowly into bigger and bigger clumps, and these bigger clumps bump and jostle one another until eventually they form "superclumps."

While they are congregating, the bumping, jostling clumps and superclumps slowly acquire angular momentum. In the case of galaxy formation, it may take many millions of years before the pull of gravity and the inertia of moving masses combine to produce rotation about a common center. The scale of the forces and distances is colossal. But basically, it's the same dynamic physics that turns the energy of random currents in a river into swirls, swirls into eddies, eddies into whirlpools. Next time you're heating water for pasta, add a couple tablespoons of olive oil to the pot. As the temperature rises and the energy level of the water molecules increases, watch how the small, floating blobs of oil churn, coalesce, begin to spin whenever they strike one another off center.

It's not much different with galaxies. Here's Peebles' somewhat more scientific summary of what happens: picture, he says "a messy blob moving away from an irregular boundary separating it from other developing protogalaxies. The unequal pull of neighboring mass concentrations on the material within a blob produces a velocity shear that in general leaves the protogalaxy with angular momentum of rotation" (Peebles 1993, 542).

Most galaxies come in the shape of spirals, discs, or wheels, strewn throughout the visible universe randomly in clusters and occasional "superclusters." Almost all galaxies of any shape spin, and (remember the chef tossing a blob of pizza dough?) as they spin, they stretch and flatten out. The rotation curve of a "flat" galaxy (as opposed to the rarer, "blob" shape) is a graphic representation of the speeds at which the stars located at different distances from the center of rotation spin around the galactic axis. The graph of a rotation curve plots the orbital velocity of an object against its radial distance. In the case of a *solid*, rotating body—a curve ball, for example, or the spinning Earth—tangential speed increases regularly with distance from the center, and the graphic representation of that data—plotted as a rotation "curve"—will be linear and rise steadily.

Orbiting planets and stars in galaxies, on the other hand, do not behave exactly like merry-go-rounds or skaters in a spoke formation. Rather, they follow Kepler's third law of motion which relates the *square* of an orbiting body's tangential speed to the *cube* of its radial distance. For orbiting bodies in space, therefore, the rotation curve normally starts high and bends *down*. This is because the more distant the planet, the slower the velocity it needs to stay in orbit. On the other hand, the closer a planet in our home solar system is to the sun, the greater the pull of gravity on it, and the faster it must move to escape spiraling inward and being burnt to a charred crisp. More distant planets in our solar system circle at a comparatively slow pace. Compared to Mercury, which chases round the sun at more than 107,000 miles an hour, Mars travels only about half that fast, while at a mere 10,700 miles an hour, distant Pluto's pace seems relatively leisurely.

In theory, such clusters of billions of stars rotating around a common center ought to behave no differently from the planets in our solar system. But they don't. Since about the middle of the twentieth century, astronomers have known two things about galaxies: (1) they spin, and (2) they don't spin the way everything else does. As early as 1939, Horace Babcock had published research data in Harvard's *Lick Observatory Bulletin* showing that the angular velocities he observed among stars in the outer parts of galaxy M31 did not conform to the expected kinematics of planetary systems. They were spinning much faster than they ought: their rotation curves went *up* steeply, then turned flat; the outer stars of disk galaxies, vastly more distant from the galactic centers, were speeding round just as fast as stars near their centers. Sometimes they spun even faster than that. Hydrogen clouds that lay far beyond the outermost visible stars of the galaxy Andromeda, for example, were, as science writer Richard Panek puts it, "spinning at a seemingly suicidal rate" (Panek 2011, 52).

One of the people who was curious about the spinning universe was a new graduate student at Cornell, Vera Cooper Rubin. By the time she enrolled in Cornell's master's degree program in astronomy, Rubin had already been looking and wondering about the stars all her life. "What fascinated me," said the astronomer in an interview for the American Institute of Physics in 1995, "was that if I opened my eyes during the night, [the stars] had all rotated around the pole. And I found that inconceivable. I was just captured" (Scoles 2016). In 1951, casting about for a subject both suitably novel and suitably limited for a master's degree thesis, Rubin wondered if it were possible to combine old theories of a mechanical, "clockwork" universe with more recent evidence of an ever-expanding cosmos. Since everything in the observable heavens was rotating—earth, sun and solar system, entire galaxies—maybe the universe was too?

By applying conventional rotational kinetics to the measurements of the galaxies' angular velocities, Rubin and her collaborators proved that these immense clusters of stars were spinning around the galactic center faster than ought to be possible. A comparison of luminosity profiles of spiral galaxies showed that overall brightness decreased (as one would expect) with increasing radial distance from the center of the galaxy (Peebles, 47). This made sense because the spiral arms of a galaxy contain fewer stars then those regions closer to the core—hence less overall mass—to light it up. But graphs of comparative velocities told a different story. They showed discrepancies between the predicted orbital speeds of stars and the stars' actual rotational speeds which remained constant even as their radial distance increased.

The combined mass of all those stars and gas clouds just didn't yield enough gravity to make the equations work; those billion points of light were orbiting at speeds sufficient to have flung themselves apart long, long ago. Something had to be holding them together. But what? The data shook Newtonian physics to its core. The Cambridge physicist's equations worked perfectly to explain the motions of the inner regions of a galaxy, but they were disturbingly inaccurate when it came to its outermost zone. There were two possible solutions to the conundrum: either Newton was wrong, or galaxies harbored huge masses of gravity-inducing stuff we just couldn't see.

The prudent choice for any physicist was to queue up behind Newton. The billions of stars contained within a galaxy couldn't possibly move at their observed speeds if they were circling around a dominant central mass, like individual planets orbiting a lone sun. Ergo, there *had* to be more matter up there than met the eye—huge amounts of matter within or around the galaxies that was pulling on the individual stars yet was somehow invisible to telescopes. If, Rubin reasoned, large amounts of mass were moved *away* from the center of galactic rotation, Newton's laws would be preserved, and gravitational forces and orbital speeds would be roughly similar everywhere. If non-luminous, or "dark" matter was strewn lavishly around a galaxy in a kind of colossal "halo," Rubin thought, overall mass would be located away from the center of rotation; the mass would be, so to speak, frontloaded.

"One day," she said, "I just decided that I had to understand what … I was looking at, and I made sketches on a piece of paper, and suddenly I understood it all" (Scoles 2016). Non-luminous material—"dark matter," as it was soon popularly termed—was conjured to exist, in short, to save the centuries-old scientific edifice that had been built on Newtonian laws of motion. "If Newtonian mechanics gives a useful approximation to the dynamics of galaxies and clusters of galaxies," says Peebles, "their masses certainly are dominated by dark matter, that is, by material with a mass-to-light ratio considerably higher than that of the matter in and around the stars seen in the central parts of galaxies" (Peebles 1993, 420). That's a physicist's explanation; in lay terms, the idea sounds more controversial. If Newton was right about the way things move, then the universe is mostly invisible.

For galaxies' unvarying radial velocities to be consistent with Newtonian mechanics, therefore, it is assumed that their anomalous, "flat" rotation curves are determined by two essentially different kinds of gravity-causing masses. Stellar velocities in the inner regions of spinning galaxies are governed largely by the mass of the stars we can see, while stars orbiting further and further from the center are accelerated by improbably huge amounts of invisible matter located somewhere in or around the outmost regions of the galaxy. Within a typical galaxy, it is currently assumed, both "light" and "dark" matter are evenly distributed to keep rotational speeds constant throughout.

Such is the consensus for now, at least. It's ironic that the biggest player in the cosmos—one might as well call dark matter the "prime mover"—is hidden from us even as it governs the motions of most of the stars in the sky. To base a cosmology on stuff we can't ever hope to see sounds more than a little dubious; a skeptic might wonder how such thinking marks an advance over Ptolemy's epicycles. To be sure, it's risky to disagree with Sir Isaac Newton, and for the time being, at least, we have little choice but to presuppose the existence of massive amounts of matter that somehow escapes our telescopes. On the other hand: when it comes to speculating about the revolutions of heavenly spheres, "it is well to bear in mind," as Peebles of Princeton dryly notes, "the alternative that we are not using the right physics" (Peebles 1993, 47).

Motorized Bacteria

On a country road in the panhandle of Idaho, my wife turns to me and asks this: Do you think if aliens saw this, they would think it was beautiful?

This is something new. My wife does not normally pose abstract questions; her engagement with the world is always more sensible. It is early afternoon, a bright October day. Yellow leaves float from aspens and cottonwood trees, tumble and swirl beside the car as we pass by. The sky is high, impossibly remote. Clouds like paper cutouts are pinned against faultless, cobalt blue.

Do you mean is there such a thing as a universal aesthetic?

No, she says, that's not what I meant. I mean whether an alien who saw this for the first time would give it the appreciation it deserves.

Nature uses rotational energy in an astonishing number of shapes and ways, whether in forming spiral galaxies or dispersing the seeds of the maple tree. But living creatures that spin or roll themselves from place to place—hedgehogs and tumbleweeds, for example—are rare. The question of why animals didn't evolve wheels, says Michael LaBarbera, "is part of the professional folklore of biology" (LaBarbera 1983, 395). As with wheels, "motorized" locomotion among animals appears to be similarly absent. In fact, it's not—but nature's motors are so rare, indeed, that they were long believed to be the stuff of science fiction.

Near the small end of nature's scale of spinning things are the fantastic, molecular "motors" belonging to bacteria such as *E. coli*. The motorized propulsion mechanism used by these ultra-primitive, prokaryotic life forms is among the most versatile in the world. Yet point for point and pound for pound, such a mechanism is better by far than legs or fins at moving those creatures lucky enough to possess it. It's better even than wings. Birds achieve motion either by using powerful muscles that connect wings to breast, or, in the case of soaring species like the wandering albatross, by taking advantage of a combination of gravity and air currents. But the great majority of birds can move only in a forward direction. (The hummingbird alone among *Aves* can hover without the aid of a headwind; it can even fly backwards for a short while.) The bacterial life forms *E. coli* and *Spirilla*, in contrast, zip around their watery environments in any direction whatsoever because the tiny, molecular machines that propel them function equally well in reverse.

That bacteria were capable of movement had been known for hundreds of years. Their mobility had been verified at least as far back as the day in 1676 when Antonie van Leeuwenhoek peered through his microscope at a drop of water and saw life on a scale so unbelievably small it had never been imagined. Before his astonished eyes, van Leeuwenhoek saw—in his words—"many thousands of living creatures ... moving among one another." At the time he first glimpsed bacteria swimming about, van Leeuwenhoek's lenses were too coarse to reveal the tiny creatures' means of locomotion. So, the Dutch scientist quite understandably reasoned that these minuscule beings were built along the lines of horses, dogs, and badgers, and were in all likelihood, therefore, "furnished with paws withal."

He was mistaken, of course. It was discovered shortly thereafter that the comings-and-goings of bacteria depended not, as van Leeuwenhoek had erroneously supposed, on sets of paws or hooves, but on the rapid, back-and-forth waving of an external, flagellar filament that functioned more along the lines of fins or flippers. But a complete picture of the bacteria's actual propulsion mechanism and its energy source began to be filled in only during the last several decades of the twentieth century (Macnab 1999). Prokaryotic bacteria have nothing in their makeup resembling the joints and bones of animals (let alone the muscles with which to move them), so they have come up with a way to achieve mobility by transferring protons across a protein membrane, thereby powering and spinning tiny, whip-like appendages located (so to speak) on their bottoms.

It's a living structure far more astonishing even than the appendages imagined by van Leeuwenhoek. These bacteria seem to have evolved a biological ensemble that looks and functions like machines engineered by humans. What's more, these fantastically small motors can spin in one direction just as readily as another. In one experiment, for example, in response to an attractant added to their environment, the bacterial motors rotated counterclockwise to move the animals toward the source, but they spun clockwise to back away from an irritant as soon as it was introduced.

Much too tiny to be visible to human eyes, these flagella are spun by reversible rotary "motors" that function according to the same electromagnetic principles that turned Peter Barlow's famous wheel. At the heart of the motor's design is something called a "flagellar motor/switch" consisting of three proteins whose names sound like a set of characters from *Through the Looking Glass*: FliG, FliM, and FliN. Of the trio, by far the most important seems to be the FliG protein because it is chiefly responsible for the bacterial motors' production of torque. FliM, on the other hand, seems to be a kind of on/off switch, while the role of FliN remains mysterious (Kihara et al. 2000).

Each bacterial motor is made up of two distinct parts remarkably like the one constructed by Moritz von Jacobi. There are differences in the construction of these motors across the many different species of flagellar bacteria, but the basic design remains consistent across almost all species: a series of stationary units ("stators") encircles a central, spinning component (the rotor). The motors obtain energy from the electrochemical potential of certain ions, typically sodium or hydrogen, which flow through the cellular membrane via a narrow, water filled "tunnel" or "channel" in the cytoplasmic membrane that allows only ions of a specific size or electric charge to pass. (Similarly, it's the electrochemical potential of sodium, potassium, and calcium ions flowing through various "channels" in your heart that keep it beating more than two and a half billion times over the course of a lifetime.) These bacteria blur the border between living creatures and man-made machines. It is, for example, an ion drive propulsion system that is theorized to be capable of accelerating future spacecraft to speeds vastly greater than are possible with chemically powered rockets.

The more "stator structures" any species of bacterium possesses, the greater its capacity for generating rotational energy, or torque, hence the greater speed with which it can move about. DNA analysis suggests that more primitive bacterial species have about a dozen stators, while speedier, more recently evolved species possess up to seventeen, leading researchers to believe that evolution once again capitalized on a novel mechanism, making it better over time at doing what it does (Science News 2018). In proportion to their size, these flagella-powered bacteria are the undisputed champions of nature when it comes to creatures that swim. Their "outboard motors" spin at phenomenally high speeds (on the order of tens of thousands of revolutions per minute), and bacteria use them to churn through their liquid environments at speeds of hundreds of body lengths per second. Admittedly those body lengths are almost too small to see—about two-thousandths of an inch, or about the half the thickness of a sheet of copy paper. But if Olympic champion swimmer Katie Ledecky were to plow through the water at a comparable rate, she would blitz the 1,500-meter freestyle event (that's thirty laps back and forth in an Olympic pool) before any of her competitors had taken their second stroke.

When I had read enough about the rear ends of bacteria to be able to have an intelligent conversation with an expert on the subject, I sought out a microbiologist at his lab in the University of Idaho. Scott Minnich has been a research biologist for decades (he obtained his PhD in 1979), and he's been studying flagellar bacteria for almost half a century, ever since graduate school when he became fascinated by these vanishingly small, motorized beings. Minnich has published dozens of scientific papers, co-authored a textbook (*Explore Evolution*), testified in Kitzmiller v. Dover (the famous "Dover Panda Trial" in which the plaintiffs argued that "intelligent design" was a form of creationism and therefore in violation of the "establishment clause" of the First Amendment to the Constitution), and he was part of a team of scientists sent to Iraq in 2004 to assist coalition military forces in assessing Iraq's biological warfare capabilities. (They had none, he and his team concluded.) Yet Minnich sees himself in humble terms, describing himself simply as an "applied scientist trying to devise mousetraps to catch these creatures." "I'm intrigued by the biology behind them," he says. "When you start dissecting them, there's an amazing technology involved."

"The flagella are propellers," Minnich explained, "with cross-sections similar to the blades of a turbine. It's a true rotary engine, built from the inside out starting with the stator. It takes about one to two percent of the individual cell's energy to build them, along with a significant amount of genetic code." (For comparison, the human brain consumes about 20 percent of the body's energy, and the heart—marvelous as it is—only about 1 percent.) "The stator is hooked into the cell wall, while the rotor is driven by proton motive force generating an electromagnetic gradient." The bacteria are entirely lacking in complex neurological structures, of course, yet somehow, they are in complete control of their motors, according to Minnich, and they are fully able to start their flagella rotating in direct response to nutrients. Amazingly, the teeny creatures can not only recognize nutrients when they come across them, they are able somehow to remember—if that is the right word—for several seconds both the location and the quality of a nutrient source. Then, when the bacteria home in on a potential meal, they motor toward it with flagella spinning at speeds up to 100,000 rpm.

The motors themselves are small beyond belief—about 40 nanometers in diameter, on average, approximately ten thousand times smaller in diameter than a human hair. "They're the most efficient machines in the universe," says Minnich, "high tech in low life…. That's the beauty of basic science; there's a whole story behind it." Indeed, bacterial motors bear an uncanny family resemblance to the array of manufactured electric motors that power everything from pencil sharpeners to Amtrak road diesel locomotives. Whether it spins a teeny flagellum or packs enough horsepower to propel a Navy warship, every electric motor is essentially crafted on the same basic design.

As with those ubiquitous electric motors that drive our mixers and lawn mowers, the ring of stators in the bacteria provide a fixed, circular force field within and against which a rotary component is made to spin. The great thing about rotary motion (as opposed, say, to linear or

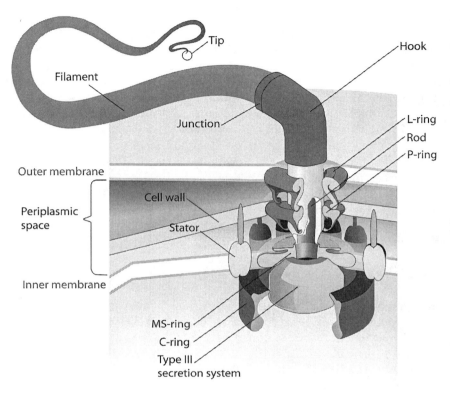

Schematic cutaway of the main structural components of a bacterial flagellum "motor." The basal body is embedded within the cell membranes and includes a stator and rotational components. The hook and filament extend outwards from the cell exterior, the filament spinning like a helical propeller (LadyofHats, Wikimedia Commons).

eccentric movements) is its directional flexibility. Unlike the architecture of wings or legs, any circular drive mechanism is essentially fully reversible. That drill in your toolbox is just as happy to back a screw out of its hole as to drive it deeper into your deck board, and it's the same with the teeny bacterial power plants. They spin equally well to the right or to the left, depending on whether the prokaryotic cell to which they're attached needs to move toward higher concentrations of chemoattractants or, alternatively, to back away from chemorepellents.

Different bacteria have developed different motors to suit their environmental needs. There's a beautiful consistency to the array of their motors. Each is perfectly adapted to a unique job and environment, just like a lineup of drills and bits on a hardware store shelf. Salmonella is that old Makita Ni-Cd rechargeable ⅜-inch drill collecting dust in your garage, while Campylobacter jejuni stands in for the DeWalt 20V Max Lithium-ion, half-inch impact driver that will sink 4-inch deck screws into treated 2 × 6's hours longer than you'll be able to wield it. Campylobacter jejuni is found worldwide in all types of birds and livestock, and it is common especially among wombats and kangaroos in Australia, the cause of so-called bushwalker's diarrhea. As few as 800 of these tiny invaders can establish a colony in a human's intestinal tract, which is why you are so often reminded to scrub your hands and counter after preparing that Thanksgiving turkey for the oven.

Bacteria like *Campylobacter jejuni* require a relatively great propulsive force (as propulsive forces are measured among bacterial species) to power through the lining of an animal's gut, and so it has twice as many stators located at greater distance from its rotor than are found in Salmonella or Vibrio cholera. From a design standpoint, it's basic rotational kinetics 101: more stators located further away from the axis of spin make available more torque for drilling through thick, sticky mucus that coats the walls of a creature's stomachs and intestines.

Researchers theorize that ancient bacteria evolved ways to reconfigure torque scaffoldings to better accommodate themselves to diverse habitats. To describe bacteria as having "scaffoldings" (the researchers' word) is accurate, but it probably fails to convey fully just how fantastic a structure these bacteria possess. We're talking an extremely small radius here—only about 45 nanometers in the case of the smallest ring of stators, with the larger "motors" more than twice that size. Another beauty of these creatures is their uncanny, entirely mechanistic design.

The language Minnich uses to describe this completely organic form as well as its functional capabilities and operation could be substituted almost word-for-word for the description of almost anything that comes out of the factories of Ford and Toyota: a "motor" applies torque to an "axial driveshaft" which transmits power first through a "universal joint" and finally to a spinning "propeller," culminating in movement. Not much different, in other words, from the power delivery mechanisms in our cars and trucks.

These tiny, spinning creatures bid fair to be the most efficient life forms in the universe. Among their most remarkable capabilities is their uncanny ability to put themselves back together again should they need repair. In this they resemble those scary, future tech forms featured in Arnold Schwarzenegger's blockbuster films *The Terminator* (1984) and *Predator* (1987). But perhaps chief among the bacteria's wonders is their apparent capacity for autogenesis. Spirilla and others like it, according to microbiologist Alan Gillen, regularly construct themselves from a disorderly heap of parts (Gillen 2015).

"A Charge that Rotates"

In January of 2019, my granddaughter, then in eighth grade, constructed a three-dimensional facsimile of a sodium atom as part of a science research project. Her model clumped 23 marshmallows into a crude ball for the atomic core—11 were stained with blue food coloring to represent protons, 12 were yellow for the neutrons. Around this "nucleus," 11 plain white marshmallows stood in for the electrons belonging to the sodium atom. The globby clumps were arranged on concentric, braided paper circles at "orbital" distances from two to six inches.

Katie was annoyed by the childishness of the assignment. She worked hard on it nevertheless, always in good humor. ("While the glue dries, I'll just sit here and eat electrons!") It wasn't her fault the project was seriously flawed from the start because the picture of the atom she had been given was wrong. Not just a little wrong, either, but hugely, staggeringly, whoppingly wrong. The proportions were *way* out of whack. In actuality, the radii of nucleus and atom differ by a factor of about 100,000; their size discrepancy alone makes them impossible to model accurately. If we could somehow peer inside an individual atom—which we can't, at least not yet—the nucleus, it has been said, would look like a pea in the middle of the Indianapolis Speedway. As for the individual electrons themselves, there's no way Katie could have represented them. In fact, *any* attempt to embody the size or location of a single electron would have been incorrect. Even to call it a "particle" is misleading; it makes us picture things like dust, sugar, salt, or the gritty bits of sand that stick to feet as one strolls along the beach. But these are all visible forms of matter. A single grain of sand takes up space: it can be measured, counted, weighed. Electrons are not like that; they're something else entirely.

Because electrons cluster about a nuclear core, it's tempting (not to mention comforting) to picture them as miniature planets spinning obediently around a central sun. This is the Ernest Rutherford/Niels Bohr model for the internal structure of an atom, first theorized in the first several decades of the twentieth century. Earlier physicists like J.J. Thomson (the discoverer of the electron) proposed that positive charges within the atom were strewn about more or less evenly throughout, like chocolate chips inside a cookie. (Thomson's version of atomic substructure was called the "plum pudding" model.) But in his laboratory at the University of Manchester, Rutherford directed a series of experiments that proved otherwise.

Rutherford and his associates, Hans Geiger and Ernest Marsden, shot a stream of heavy alpha particles at a patch of gold leaf and recorded the directions that the particles rebounded, or "scattered," after they hit the target. The results at first seemed implausible. Almost all the particles buzzed straight through the solid metal, apparently hitting nothing; it was like trying to shoot a goldfish in a barrel the size of a container ship. Odder still: now and then, particles careened wildly off course, like a running back bouncing off a tackle. And on a few singular occasions, they boomeranged directly back at the source. "It was then," Rutherford wrote later, "that I had the idea of an atom with a minute massive centre, carrying a charge" (Needham and Pagel 1938, 61). Rutherford's proof was irrefutable: an atom was mostly empty space, nothing at all—although somewhere deep inside each was a tiny, dense core. Even more astonishing: the many different qualities we attribute to everyday objects—hardness, depth, a glossy surface—were mere illusions, a nothing that is, a nothing that always was.

Rutherford initially believed electrons coupled in rings with positive charges in the atomic nucleus, much as in Katie's clustered marshmallows. Then, using this planetary model of atomic structure, Bohr was able to apply classical rotational kinetics to calculate both the angular speed of an electron circling the nucleus of a hydrogen atom (about 8 million kilometers an hour) as well as the radius of its inmost orbit—five hundred billionths of a meter (0.0000000000529m). Not that anyone could hope to see this happening, of course. Even assuming we could spot something that small, it would flash by well before eye and brain could pick it up. Yet the Rutherford/Bohr model for sub-atomic structure was powerfully attractive. Its beauty lay in its consistency, its mirror-like repetition of the basic design of the universe: top to bottom, impossibly big or unimaginably small, the cosmos was uniform all the way down. Picturing the atom as a miniature solar system instantly kindled the public imagination. Images of a nuclear "sun" surrounded by a cluster of orbiting, planet-like electrons were soon adopted as the logo of groups ranging from the United States Atomic Energy Commission to the Albuquerque Isotopes (a minor league baseball team), to the society of American Atheists.

Electrons are peripatetic; within an atom, like small children clambering over a merry-go-round, electrons can transition (or "jump") from "higher" to "lower" orbital states, and vice versa. When electrons undergo these transpositions, energy is always given up or taken on. This is because their energy, like angular momentum, is a conserved quantity; it remains constant. Just as an acrobatic diver leaping from the high board spends the potential energy that was accumulated in climbing to the top of the platform, so electrons cannot move from one state (or "level," "orbit," or "shell") to another without at the same time gaining or giving up a small amount of energy.

For a diver, the quantity of energy spent on the climb and expended in the plunge can vary depending on factors such as weight or the height of the board above the pool. It's not so with an electron, however, because with every transit from one state to another, an electron gives up—or takes on—only certain, fixed amounts of energy. The energy emitted or absorbed by electrons, in other words, is "quantized"; it comes in packets, or specific amounts. You can think of the transpositions of electrons as if you were to go shopping with a pocket full of twenty-dollar bills, at stores that give no other denominations in change; you can make transactions only in twenties or its multiples. In the case of electrons, those indivisible multiples of energy are called *photons* (Greek φως, *phos*).

One method of keeping track of the emission or absorption of photons by electrons in transition is with a spectroscope, which is basically a kind of prism. It was with a spectroscope, for example, that Newton identified seven separate colors that together made up all the light we see.[2] Spectroscopic analysis goes one step further in that it allows for even subtler investigations of matter. It's basically a snapshot of photons as they are emitted—or absorbed—when electrons move about inside a particular substance. When sodium atoms (as in Katie's model) are heated, for example, the electrons in them absorb some of that thermal energy and transition to a "higher," more "excited" state or "orbit"; as the atoms cool, the electrons fall back down to a more quiescent level. Within each atom, electrons move readily back and forth between the different, discrete energy levels (or "orbits," or "shells") much as a collection of tipsy conventioneers might ride the hotel elevator up and down, stopping on different floors.

But heated electrons don't stay buzzed for very long—like partygoers coming off a binge, they soon fall back to their normal energy levels, and when they do they give off photons that can be seen using a spectroscope. These bits of light energy appear as individual lines arranged across a spectrum. They can be seen, says Jon Butterworth of University College of London, "either as dark lines in the spectrum of light passing through a material, where those wavelengths of photon have been absorbed, or as bright lines in the light given off by a material when it is heated, such as the characteristic yellow lines in a sodium lamp" (Butterworth 2018, 51).

There's more: because each different element, whether carbon, sodium, or manganese, has its own unique configuration of electrons that makes it different from every other element, when any material is heated to incandescence, as it gains or loses energy, it will emit—or absorb—its own, unique array of photons. Individual elements leave behind these unique, identifying traces much as Parisians once dropped off personal *cartes de visite* in the households of their friends; whenever the owners returned home, they could check the bowl of *cartes* in the vestibule to see who had earlier stopped by. Every atom, ion, or molecule anywhere in the universe similarly displays its own elemental, identifying spectroscopic "signature" or "fingerprint." Spectroscopy is an extraordinarily useful way to determine the chemical composition of a substance; in fact, it is only through spectroscopy that we can tell what the things beyond our world are made of, that the remotest stars in the heavens are made of the same stuff as you or I.

During the first two decades of the twentieth century, physicists were increasingly puzzled by anomalies—"splittings," as they were called—that appeared in the spectroscopic images of atoms that were immersed within a magnetic field. Pieter Zeeman (who shared the 1902 Nobel Prize in physics with Hendrik Lorentz), for example, had showed that the spectral signatures of sodium and cadmium *changed* whenever they were exposed to magnetic fields. What was without magnetism a unique spectroscopic picture of bright, single lines became blurred multiples—"splittings"—of two or more transverse or longitudinal streaks. That the same element should come so to speak with different calling cards seemed contradictory; an element ought to be, well, elemental.

Lorentz suggested that the anomalies might be explained if it were assumed that when electrons inside an atom were in the grip of a magnetic field, they commenced vibrating, thus blurring the spectroscopic lines as electrons took on or emitted energy. Lorentz's physics seemed to work for the patterns of simple elements, but soon other, more complex splittings were recorded that disproved his supposition. Even Niels Bohr's then revolutionary model of the atom, the first to assign electrons to distinct orbital "shells," proved unable to account for the splitting phenomenon. Zeeman's response to the mystery he had uncovered was as classy as it was humble: "Nature gives us all, including Professor Lorentz, surprises."

A partial glimpse into the contradictory behavior of electrons came in 1919, when the young German American physicist Alfred Landé (in his own unabashed recollection) "cracked the magnetic code of atomic structure." Landé proposed that electrons don't just simply *react* to a magnetic field (as Oersted's electrified wire had twitched when a magnet was moved near to it). Instead, Landé suggested, electrons also *carried* an inherent magnetism, something he called "the g-factor," a field force which could account for their increased volatility when subjected to additional magnetic influences (Hentschel 2009). Where could this supplemental magnetic force come from? The same source, Landé guessed, that produced magnetic effects for Hans Christian Oersted when he first twiddled a live wire back and forth in the vicinity of a compass. The source of the electric current—the electron—had to be in motion.

Among the physicists who became curious about whether electrons might somehow be generating magnetic fields with their movements was a young physics student at Columbia University named Ralph de Laer Kronig. In January of 1925, Kronig was in Germany listening to a lecture by the renowned theoretical physicist Wolfgang Pauli. Kronig was then barely twenty years old. But he had taken the advice of another young physicist, Paul Ehrenfest, and interrupted his studies at Columbia to visit Germany and Copenhagen to take advantage of the ground-breaking research then being conducted there on theoretical physics.

A few months earlier, Pauli had speculated that the electrons in certain alkali metals seemed strangely duplicitous: that is, they had an either/or characteristic. (Pauli's term for this doubling, "an indescribable two-valuedness," seems to verge on mysticism.)[3] Kronig met with Landé, who showed him a letter from Pauli which referred to electrons' duplicity. Kronig thought that this Jekyll/Hyde nature of electrons could be explained if they were assumed to possess small amounts of angular momentum—if, that is, they were spinning like teeny, tiny tops. When they first heard of a spinning electron, the founders of quantum theory, Walter Heisenberg and Pauli, dismissed it out of hand. Pauli all but guffawed; he conceded that Kronig's idea was "very clever," but added that it "has nothing to do with reality."

Kronig got to laugh himself only a short while later, as it turned out. The following year, Sam Goudsmit, a graduate student at Leiden University in the Netherlands, had been playing with the mathematics of the hydrogen atom using only *half* the normal quantum values for electrons rather than the conventional whole integer. This gave electrons the option, so to speak, of occupying in-between states. (Picture an elevator somehow taking on and dropping off passengers halfway between floors.) The results of Goudsmit's calculations were stunning—it was (his words) "like magic." Goudsmit shared his work with his lab partner, George Uhlenbeck; immediately he, too, grew excited. It was one of those crazy, once-in-a-lifetime moments of insight; even decades later, Goudsmit was able to recall their conversation with preternatural clarity: He [Uhlenbeck] said to me: "[b]ut don't you see what this implies? It means that there is a fourth degree of freedom for the electron. *It means that the electron has a spin, that it rotates."*

The two young researchers rushed with their findings to Paul Ehrenfest, in whose lab they had been employed. Ehrenfest promptly (according to Goudsmit) told them that their work was wrong. "But send it off for publication anyway," he advised—almost certainly with his tongue in his cheek. "What could be the harm?" he teased. "You have no reputations to start with, so you have nothing to lose" (Goudsmit 1971). Ehrenfest's advice sounded specious, to say the least. Goudsmit was wary; he wanted to run more tests, work through more sets of calculations. But Uhlenbeck was unable to restrain his enthusiasm. Off went the article for publication.

The idea of a spinning electron indeed seemed preposterous—how could you test such a notion, let alone see it happening?—and at first it was treated as such. A spinning electron, as Max Born once put it, was "not to be taken literally" (Born 1935). Once more, Pauli scoffed; some quick calculations, he said, told him the surface of such an electron, were it to exist, would have to be spinning faster than the speed of light to generate the angular momentum it seemed to possess. For a time, the idea of electron spin lay undeveloped. But within a few years some important names got on board the spin train. Even Pauli—who had once been adamant that Goudsmit's and Uhlenbeck's spin theory, because of its classical, mechanical character, could not be valid within the quantum *Weltanschauung*—finally conceded that electrons somehow possessed an intrinsic angular momentum along with a corresponding magnetic moment. In all likelihood, Pauli thought, the particles were not really spinning—who, indeed, could ever tell whether they were? But it certainly seemed *as if* they were spinning, and that was good enough[4] (Giulini 2008).

By postulating infinitesimally small bits of matter that were in effect spinning furiously—more accurately, by penciling inside an atom particles that carried intrinsic, fixed amounts of angular momentum—the inner dynamics of an atom began to make sense. Electron spin made it possible for two otherwise identical, electrically repulsive charged particles to coexist within the same orbit or shell. Electrons (and quarks, the "seeds" of protons and neutrons) have properties that can best be defined in terms of rotational energies, or angular momentum. On the model of a lodestone with opposite "north" and "south" poles, some carry half an "up-spin," while others were assigned a half "down." That's where things have come to rest, though the picture is far from clear. Quantum spin is still a "mysterious beast," says Takeshi Oka, emeritus professor of astrochemistry at the University of Chicago, "yet its practical effect prevails over the whole of science. The existence of spin, and the statistics associated with it, is the most subtle and ingenious design of Nature—without it the whole universe would collapse"[5] (Tomonaga 1997, viii).

Why is it "spin-up" and "spin-down," though, rather than "left" and "right," clockwise and counter-clockwise, or some other pair of opposites? The answer, ironically, also comes straight out of classical mechanics. Spin—angular momentum—is a vector quantity; it must, therefore, have a specific direction. But the tangential direction of a spinning object is always changing. The only unique direction associated with it is along (or parallel to) the axis of rotation; by convention, then, that's the spin vector. In the everyday world, a "right hand rule" assigns variable, either positive or negative, values to the spin vector depending on its angular velocity and direction of rotation.[6] In the world of subatomic particles, things are different. Electrons are assigned quantized, half-units of spin, either one-half spin-up or one-half spin-down; a matched pair (or pairs) of electrons, one spin-up and one spin-down, can occupy one or more of the different energy levels inside an atom. These different energy levels are sometimes called electron shells, or "orbits."

But it's almost certainly wrong to picture electrons spinning like infinitesimally small planets circling an infinitesimally small, nuclear "sun." "Spin" in this instance refers not to actual, physical rotation but to an intrinsic quantity of angular momentum. In the same way that theorizing electromagnetic "fields" of force allows us to see that electricity and magnetism are in some way identical, so theorizing the "spin" of subatomic particles allows us to grasp the reality of things we can't see in terms of things we can. Electrons can be understood as having two interrelated fields, therefore: an electric field because of their negative charge, and a magnetic field because of their "spin"—much like the electromagnetic landscape of Barlow's wheel.

The spin of electrons differs from the spinning of objects in the everyday world, however, and it's necessary to picture their "spin" in some completely different way. Baseballs, gyroscopes, skaters, and bullets have mass and dimensions. Such objects take up a fixed space; they are *here, there, somewhere.* Not so with electrons, however, which cannot be viewed as localized objects, much less as literally rotating. Electrons exist in two distinct states that owing to some intrinsic property makes them behave like infinitesimally small, magnetized loops of electric current carrying intrinsic, opposite quantities of angular momentum.

Physicists universally call this property "spin," even though for all intents and purposes electrons are massless and lack a center about which to rotate.[7] It would be meaningless, therefore, to picture their angular momentum as being the result of their rotation about an axis. Yet electrons and certain other elementary particles seem to behave *as if* they were spinning, and nobody knows why. "It is as if," writes John Maddox, editor emeritus of *Nature* magazine, "space has a previously unsuspected property that requires any pointlike object such as a particle to carry a certain amount of angular momentum" (Maddox 1999).

Even this account may be misleading. Objects such as electrons cannot be imagined as "particles" in the ordinary sense of that term, as if the only difference between subatomic bits of matter and marshmallows is that the former are much, much smaller. The constituent pieces of atoms are more accurately thought of as essentially dimensionless foci of electrical charge that are capable of interacting with a myriad of other, similarly charged but otherwise featureless foci, or "points." About all we know with certainty is that small groups of such negatively charged, dimensionless points hang out with individual, positively charged clusters of somewhat more massive points. These various, electrically charged groupings form the basic atomic elements that make up everything of any substance in the universe, bacteria on up to stars. Individual electrons belonging to a single atom, however, are not to be imagined as golf balls buzzing here and there in an otherwise empty Wal-Mart. In fact, it's best not to picture electrons in terms of their size at all. "It is not in the premise," as the poet Wallace Stevens once said, writing about an ordinary evening in New Haven, "that reality is a solid." The British physicist James Jeans, speculating about equally ordinary events inside an atom, comes to much the same conclusion: "a hard sphere takes up a very definite amount of room; an electron—well, it is probably as meaningless to discuss how much room an electron takes up as it is to discuss how much room a fear, an anxiety, or an uncertainty takes up" (Barnett 2018).

So what do electrons look like, somehow spinning inside an atom? It's hard—probably impossible—to say. As with their size, speculation about the appearance of electrons is useless. Our eyes have evolved to be sensitive to wavelengths of about 400 to 700 nanometers—that is, between 4 and 7 millionths of a meter. The approximate diameters of atoms are estimated to be about a thousand times smaller than the wavelength of visible light, which means that even if individual atoms and their constituent parts were as hard and bounded as billiard balls— and they're not—we still couldn't see them. At least, we couldn't see them in the ordinary sense of the term.[8] The waves of ordinary light are so large in comparison with an object the size of an atom, they simply can't detect that anything is there. The scene inside the atom is impossible to draw, and to picture the "spin" of an electron, as Percy Shelley said of love, beauty, and delight, is a task that "exceeds our organs, which endure/No light, being themselves obscure."

Coda

James Clerk Maxwell and
the Defenestration of Cats

Oh, and one more thing: about that falling cat I mentioned a hundred pages or so ago. The problem is this: how is a cat dropped upside down able to spin around in mid-air and land on its feet? The animals seem not to care about the principle of the conservation of angular momentum, which holds that something that isn't rotating to start with needs an external push of some kind to start it spinning. Expressed more precisely: in the absence of external force, how can a freely falling body *acquire* angular momentum when it possesses none to start with? Believe it or not, many very smart people have puzzled for years over why cats always land right side up. Their roll includes scientific luminaries like James Clerk Maxwell of the famous "Maxwell's Equations" (the series of mathematical formulas that inspired Einstein to develop the theory of special relativity). When young Maxwell was a student at Trinity College in Cambridge in the 1850s, he made a habit of throwing cats out windows just to see if he could figure out how the animals did it.

Over the centuries, various solutions have been proposed to explain how cats perform their magic. All were wrong. For example, it was sometimes argued that the cat gave itself an initial rotation by using its legs to push off from the holder's hands just as it was released, much as an acrobatic diver gains angular momentum for airborne somersaults and twists by shoving off from the edge of a springboard. But the advent of high-speed cameras in the late nineteenth century proved conclusively that cats began to rotate only *after* they were released. (At about the same historical moment, high-speed camera work answered decisively another, somewhat more famous, conundrum regarding the motions of animals: whether horses at full gallop ever had all four feet airborne at the same time. As Edward Muybridge's 1878 stop-motion movie proved conclusively, they did!)

As I'm writing this, the falling cat problem *still* has not been fully settled. But here's what current researchers think. The explanation will be a little clearer if you think of the animal as consisting of two interconnected cylinders, one up front (the head and shoulders) and a second half consisting of the hind quarters and tail. Each body half, or cylinder, has its own moment of inertia (call these I and I'). As soon as the cat starts to fall, it quickly draws in its two front paws, like a skater pulling in her arms. This makes I temporarily less than I'. Next the animal turns its tucked-in front paws a little ways round, creating an internal torque on that upper half of its body/cylinder. Since the total angular momentum of the entire system must be "conserved" at net zero (which is where it was to start with), this initial movement of the upper cylinder (I) must be balanced by an involuntary rotation of the rear torso and legs (I') in an opposite direction.

Next the animal simultaneously contracts its *back* legs and extends its forelegs, then gives its hind part a turn. This movement initiates (as before) a similar, compensating rotation on the other body-cylinder, and now it's the front half that turns. As it falls, the cat repeats these motions, alternately tucking and extending its fore- and hind legs and twisting front and back halves of its body as often as is necessary to complete a half-turn and land gracefully on its feet. At no time while it is falling does the animal add any angular momentum to the system; the cat spins, *and* the laws of physics are preserved (Essen and Nordmark 2018).

That's a brief rundown of contemporary physicists mulling over the problem of falling cats. As for when or how cats learn to do this? Now, there's a question not likely to be resolved by any known science of spin!

Chapter Notes

Preface

1. I was not the first kid to be spooked by the sensation of angular momentum being conserved. A publication dating from 1897 blithely entitled *Every Boy's Book of Sport and Pastime* noted of the gyroscope "the tendency of the thing to move ... as if it were something alive" (Clifford Pickover, *The Physics Book*, p. 218).

Chapter 1

1. See, for example, Harold Crabtree, *An Elementary Treatment of the Theory of Spinning Tops and Gyroscopic Motion* (London: Longmans, Green, 1909), and Péter Hantz and Zsolt I Lázár, "Precession Intuitively Explained," *Frontiers in Physics*, frobntiersin.org, February 8, 2019, accessed 10/23/2020).

Chapter 2

1. The greens are not being moved outward by what is popularly called centrifugal "force"; rather they are undergoing constant radial acceleration which is a change in their velocity vector. The faster you crank the handle, the greater the leaves are accelerated, therefore the greater the net force the walls of the bowl will exert against them.

2. Rolling friction is almost always *much* less than sliding friction. For example: to slide a hundred-pound block of wood across the floor you would have to keep pushing it sideways with more than sixty pounds of force; the coefficient of friction of oak on oak is about 0.62. But as a young man I could push my 2,000-pound car down the street to jump start it when the battery was dead; once the vehicle was rolling, the coefficient of friction between tires and road was a measly 0.015, so it took only about 30 pounds of pushing to keep it moving.

3. Some credit "John Reed" with the invention of the rolling pin. Since women historically identified themselves only by initials to obscure their sex, it's impossible to know whether we owe the modern rolling pin to Judy or John.

4. Modern gas and electric dryers also spin clothes inside a large drum, but the drums in these appliances, by design, spin so slowly that centrifuge effects on the damp clothes are negligible if not nonexistent. A dryer spins slow enough to keep socks and shirts tumbling freely and loosely inside the drum so they are constantly exposed to hot, circulating air.

5. Pissing on clothes to clean them may sound gross, but it's not dumb. Urine breaks down into ammonia, which happens to be an excellent cleansing agent. Ammonia is especially good at "whitening whites and softening your bath towels," according to Amanda Bell, who says also that it "makes stains disappear with minimal effort" (Amanda Bell, "How to Use Ammonia in the Laundry," hunker. com, accessed 9/10/2020).

6. The formula is $\tau = r \times F$, where τ is the torque vector and r the distance from the center of rotation to the place where F (the force) is being applied.

7. The earliest such brace isn't much older, dating from Greifswald around 1370.

8. The opposite motions are true for most Japanese woodworking handsaws, whose teeth are sharpened on the rear to cut on the back stroke. At first this sounds silly, as it's obvious that one can exert much greater force pushing a saw blade forward and down rather than backwards and up. But no, that's only half the story. Because a Japanese saw blade is under tension rather than compression on the cutting stroke, it can be made of much thinner metal and so cut with greater finesse. I've used such a saw a couple of times; it produces an amazingly smooth cut, which makes it ideal, for example, for cabinetmakers.

9. The so-called spinner shark (*Carcharhinus brevipinna*) in fact rotates rapidly on its longitudinal axis while feeding, snapping wildly as it speeds upward, spinning through a school of fish.

Chapter 3

1. Some complain that Rumi's poetry is most often read stripped of its Muslim content. In a largely secular culture such as ours, this is perhaps to be expected: the poetry of Chaucer and Tennyson is deeply Christian, a feature now not nearly as important to readers as once it was. See Rozina Ali, "The Erasure of Islam from the Poetry of Rumi," *The New Yorker*, January 5, 2017, newyorker.com, accessed 9/3/2020.

2. *The Journal of Physical Therapy* reports on a group of several dozen undergraduates who were tested for postural sway. Asked simply to stand still for half a minute, more than half of the subjects swayed involuntarily more than an inch back and forth, left and right (Sivakumar Ramachandran, "Measurement of Postural Sway with a Sway-Meter," *Journal of Physical Therapy*, January 2010, researchgate.net, accessed 9/2/2020).

3. The formula is simple: take the moment of inertia of a rotating mass and multiply it by its angular velocity. $L = I\omega$, where L (angular momentum) is defined as the product of the moment of inertia of a rotating object multiplied by its angular velocity.

Chapter 4

1. Robert Adair, a physicist from Yale, sums up the two points-of-view: "Does the curve ball then travel in a smooth arc like the arc of a circle? Yes. Does the ball 'break' as it nears the plate? Yes. Neither the smooth arc nor the break is an illusion; they are different descriptions of the same reality" (*The Physics of Baseball*, New York: HarperCollins, 1990, p. 50).

2. Air resistance (or drag) affects a baseball to a much greater extent than is generally appreciated. Robert Adair calculates that a ball hit 400 feet at the old Shea Stadium in Queens would sail almost twice as far (750 feet) in a vacuum (Adair, *The Physics of Baseball*, New York: HarperCollins, 1990, p. 10).

3. In fact, as the air molecules swarm and rush around and over the baseball, they will tend to accelerate and so exert somewhat less pressure on the boundary layer, allowing it to cling more securely to the surface. This sounds odd; but you will probably remember from high school learning about Bernoulli's principle, which stipulates the inverse relationship between the pressure exerted by a fluid and the speed of its flow.

4. The back-spinning fastball seems to rise, but it does not; it does, however, present the *image* of rising or "hopping" because the forces acting on it resist the pull of gravity in a way contrary to what the batter expects. As with the downward "break" of a curve ball, the last-minute "hop" of a good fastball comes at precisely the moment gravity ought to be accelerating the ball more steeply downward. This also explains why home-run hitters typically undercut the ball to impart backspin that helps it stay aloft and travel farther.

5. Check out this video to see what happens to a dove when it's struck accidentally by a 95+mph fastball thrown by Randy Johnson: https://www.bing.com/videos/search?q=randy+johnson+bird&&view=detail&mid=2EA5A39009FF013F7BCC2EA5A39009FF013F7BCC&&FORM=VRDGAR.

6. From a pitcher's point of view, the best knuckleballs make from one-quarter to one-half revolution before passing over the plate (or somewhere in the vicinity of

the plate). The partial turnover ensures that the orientation of the ball with respect to the air and the batter is always changing ever-so-slightly, thereby inducing even more unpredictable variations in the ball's flight path from beginning to end.

7. Serious baseball analysts refer to something called "total spin" which sums together motion in three different directions of rotation: top (or backspin), sidespin, and something called "rifle-spin" which is rotation along an axis perpendicular to the line of flight. Since true "rifle spin" does not add significantly to a baseball's "break" or "hop," it's typically subtracted from a pitcher's overall spin metrics (Max Goder-Reiser and Julia Prusaczyk, "Comparing the Rapsodo Baseball Device to Other Pitch Trackers," *The Hardball Times*, fangraphs.com, January 10, 2017. Accessed 1/6/2019).

8. Golf balls have dimples, and sharks have sandpaper skin: the roughened surfaces improve both balls' and animals' efficiency in moving through a fluid medium, changing the way air or water flows around them. As the shark undulates back and forth while swimming, its rough skin causes a vortex to form in front of the animal, sucking it forward as if it were a cyclist riding just behind a speeding van. Johannes Oeffner, a research assistant in Harvard's Department of Biology, has calculated that a shark's rough skin improves its swimming efficiency by about 12 percent (David L. Hu, *How to Walk on Water and Climb Up Walls*, Princeton UP, 2018, p. 108).

Chapter 5

1. You can spot the PIE familial resemblance between the reconstructed, prehistoric word *kʷekʷlos* and Greek κύκλος, "cycle." Because a related PIE stem *kʷel-* meant "to turn," according to Anthony, it's likely that *kʷekʷlos* meant specifically "a thing that turned"—a wheel, that is to say—rather than the more abstract concept, "circle." Anthony says also that the speakers of PIE seem to have made up their own word for wheel rather than borrow a foreign term, strongly indicating a native (and not, therefore, imported) origin of the device.

2. There are two long-standing and inter-related debates about humans, nature, and wheels: first, whether the wheel is a purely cultural achievement that was "nature-independent," and second, why natural selection did not lead to the evolution of creatures that moved about on wheels. With reference to the former question, Gerhard Scholtz argues that Egyptians may have been inspired to use wheels for transportation by observing the scarab beetle shaping dung into large balls to facilitate moving them about ("Scarab Beetles at the Interface of Wheel Invention in Nature and Culture?" *Contributions to Zoology* 77 (2008), http://www.ctopz.nl/vol77/nr03/a01, accessed 10/9/2019). As for speculations on the lack of wheeled creatures in evolutionary history, see Natalie Walchover, "Why Don't Any Animals Have Wheels?" livescience.com, August 6, 2012, accessed 1/21/2020.

3. There's even a wheel large enough to take passenger boats for a spin, the 115-foot-tall Falkirk Wheel that was built to swing boats from the Union Canal to the Forth and Clyde Canal. Designed to look like a cross between a propeller and a whale's ribcage, the Falkirk Wheel slings more than 500 tons of boats and water as effortlessly as Ferris's wheel spun fair-goers.

4. Capital punishment was first carried out using an "electric chair" in 1890. The execution of William Kemmler, who had been sentenced to death for the murder of the woman he was living with, lasted for eight awful minutes. By the time Kemmler finally died, according to the *New York Times*, the stench coming from his bleeding, horribly singed body was unbearable (Richard Cavendish, "The First Execution by Electric Chair," historytoday.com, August 2015, accessed 7/8/2020).

5. So named after Saint Catherine of Alexandria (c. 287–305) who, according to legend, was executed on a large, spiked wheel that miraculously broke in pieces when she touched it. During the centuries after Saint Catherine was broken on the wheel, the device that bears her name has acquired multiple, bewilderingly

benign applications: a British rock band, a spoked circular window, a gymnast's cartwheel, and a spinning firework also share the name "Catherine Wheel."

Chapter 6

1. This is why from early on, long-barreled guns with grooved bores were called "rifles," to distinguish them from shotguns, muskets, and other kinds of "smooth-bore" firearms. Nowadays both pistols and hunting rifles are manufactured with rifling; only shotguns and historic replicas are made as smooth-bore weapons. Some smooth-bore, short-barreled firearms are illegal under the 1934 National Firearms Act because they can be so readily concealed; if you want such a gun, you've got to submit a request to the Bureau of ATF. It was allegedly a violation of this law that started Randy Weaver of Naples, Idaho, on the road to tragedy.

2. Here's a summary of the test results. Of a total of two dozen arrows with off-set fletching shot from a range of about 40 feet, half (12) struck within the central, 3-inch ring (the so-called 10-ring) on a standard archery target. This was a slightly better score than was achieved by the arrows with straight fletching; only one-third (8) of those hit the 10-ring. But that wasn't the most interesting outcome of the test. Of greater practical significance for would-be marksmen was that *all* 48 arrows, both straight- and offset-fletched, hit the target within the next, slightly larger 9-ring. In other words, we're talking here about a difference of a couple of inches at most, not country miles or broadsides of barns. In the forests or on the battlefields of Europe during the twelfth and thirteenth centuries, straight fletching would have offered a degree of accuracy more than adequate to bring down a stag or enemy soldier. See "Helical vs. Straight Fletch: Accuracy and Repeatability," archeryreport.com, accessed 11/30/2020.

3. Magnus conducted his experiments fully aware of Robins' groundbreaking work. Indeed, the "Magnus" force in some older scientific journals is referred to as a "Robins' force," much as *crème*

brûlée is occasionally called "burnt cream" by stalwart Brits who insist that the invention of the famous custard dessert took place in the kitchens of Trinity College at Cambridge.

4. In *Custer's Gatling Guns*, Donald Myers, a former Marine colonel, speculates what might have happened at the Battle of the Little Big Horn had the impulsive, yellow-haired cavalry officer not left his Gatling guns parked a few miles away on the banks of the Yellowstone River (Donald F. Myers, *Custer's Gatling Guns*, CBC Publishing, 2008). In Custer's defense, see C. Lee Noyes, "The Guns Custer Left Behind Would Have Been a Burden," historynet.com, accessed 4/11/2020.

5. So named after Gaspard-Gustave de Coriolis, who first described it mathematically in a paper published in 1835; we'll learn a lot more about this crucial natural force (strictly speaking, in Newtonian physics it's not a force but a "pseudoforce") in a subsequent chapter.

6. When aircraft "break" the sound barrier, they create a loud "sonic boom." Supersonic bullets also make the same kind of noise, separate from the "bang" caused by escaping gases and audible downrange as a sharp "crack."

Interchapter II

1. From Old English *lād*, or "way": the dark power inherent in a lodestone was used by mariners to make a compass, or "way-stone."

2. The unvarying torque output of electric motors is responsible for a phenomenon that puzzled me as a child: why did steam locomotives, but not diesels, always back up before departing the station with their trains? The answer, as I learned while writing this book, has to do with the different ways each type of engine makes use of spin. Diesel locomotives in the middle decades of the twentieth century were *less* powerful than the steam-driven engines they were built to replace. The largest steam engine built in 1941 for the Chesapeake and Ohio Railway was rated at 7,000 horsepower, almost twice that of a then contemporary diesel locomotive. But C & O's behemoth

produced power over a torque curve that increased in linear proportion with speed. Backing up before starting, therefore, bunched the train and thus allowed the steam locomotive to get its load moving one car at a time, steadily increasing power to the driving wheels as they gained velocity. The combustion engine housed in a "diesel" locomotive, on the other hand, functioned only to spin large electric motors housed between the trucks, and these electric motors easily provided more than enough tractive force to pull the entire train away from a standstill all at once. The difference in the torque output curves between the two different sources of locomotion gave rise to a witticism once common among railroad workers: "A steam engine can pull a train it can't start, and a diesel can start a train it can't pull."

Chapter 7

1. But only up to a point: above a critical angle the airflow over the wing will suddenly peel away from the surface. When this happens, pressure differential and downwash largely disappear, and drag greatly increases. In this condition the wing is said to be "stalled," and the airplane (or bird) falls from the sky. See Alexander, *Nature's Flyers*, pp. 26–28.

2. A helicopter combines both these functions, lift and propulsion, into a single set of rotating blades/wings. To move forward, the pilot pitches the nose of the aircraft slightly downward, adding a horizontal vector component to the lift. A flapping wing also rotates around the bird's shoulder pivot, just not in a complete circle.

3. Though I didn't know it at the time, I was then looking at the world's first municipal wind farm. (Penny Farmer, *Wind Energy 1975–1985: A Bibliography*, Springer, 1986, p. 142). Decades later, stationary blades dangling from dysfunctional turbines, Livingston's futuristic towers had become an eyesore, prompting state legislator Jim Keane to write a bill requiring wind farm owners to post bonds to ensure decommissioning and removal of their giant machines at the end of their useful life (Karl Puckett,

"Wind Farms Put Up Bonds to Ensure Towers Are Taken Down," *Great Falls Tribune*, November 9, 2017, greatfallstribune.com, accessed 6/21/2020).

4. It's a small world. By the time my son Rob stopped building competition model aircraft, the balsa wood he used in their construction had become increasingly expensive, and the best wood—that would be the lightest—was unavailable. Much of the world's supply of balsa logs was being bought up by manufacturers of the blades of wind turbines.

Chapter 8

1. In 2018, analysis of several tornadoes that developed in Kansas and Oklahoma indicated that their whirling winds seemed to rise from the ground up and not the other way around! The findings contradict decades-old understanding of how these violent storms form. See Katherine Kornei, "Surprise! Tornadoes Form from the Ground Up," sciencemag.org, December 13, 2018, accessed 1/19/2020.

2. This supra intelligence was never imagined by Laplace to be human. Even if humans could never know all things at all times, Laplace explicitly denied any role in determining the course of events to chance, accident, or what Aristotle once called ἁμαρτία, *hamartia,* a "missing of the mark," or colloquially, a fuck-up.

3. The widespread term for such storms in the northern hemisphere is not cyclone but "hurricane," from *juracán*, which is what such a storm was called in the language of the Taino, a people who inhabited many of the Caribbean islands at the time of their discovery by European voyagers in the fifteenth century. In calling Atlantic cyclones by their original name of "hurricanes," therefore, we are unconsciously memorializing a long vanished, long mistreated "first people"—not the only instance in history of this sort of backhanded linguistic tribute.

4. During the writing of this book, I had two college physics textbooks constantly at my side. Only one gave any account of the Coriolis Force, and that description, strangely, was not included in the several chapters devoted to rotational kinetics. An explanation of the

Coriolis Force appeared only in an appendix, along with basic reviews of algebra, trigonometry, Gauss's law, and Galilean transformation equations. In the other book, Coriolis and his "force" received no mention whatsoever.

5. The term "Coriolis Effect" is often used simultaneously with the "Coriolis Force" or even as a substitute for it. This may lead to some confusion, especially when comparing different reference sources. (It certainly confused me!) A Coriolis "force" (call it a "pseudoforce," if you like), acts on any object moving freely in a rotational frame in direct proportion to the rotational velocity *of* that frame. The further an object is from the axis of rotation, therefore, the greater the Coriolis Force on it according to the formula: $a$cor = $2\omega\upsilon$, where a is the Coriolis acceleration, ω the angular velocity of the rotational frame measured in radians, and υ the tangential velocity in a plane measured perpendicular to the axis of rotation, i.e., sideways. On the surface of the earth, therefore, the situation is the same as with the boy and girl playing catch on the rotating crop circle. This sideways vector quantity ($a$cor) will obviously be greatest at the equator, where an object has a rotational speed of approximately 1000 miles an hour, and it will steadily decrease to near zero at latitudes progressively closer to the poles. But the Coriolis "effect"—that is, the amount of sideways deflection experienced as an object moves in a northerly or southerly direction—will appear to be greatest at the poles and diminish progressively until it approaches zero at the equator.

6. A common misconception is that the Coriolis force causes waters disappearing down a toilet bowl to swirl in one direction in the northern hemisphere, the opposite way south of the equator. When my wife and I moved to Australia in 1969, we were keen to observe this phenomenon. By the time we arrived in Melbourne, however, we couldn't remember with confidence which way the disappearing waters had turned back home in Philadelphia. Not that it mattered, however. We later read up on the subject and learned that the difference between the east/west rotational velocities of the

different sides of a toilet bowl is so small it's of no consequence in comparison with other, mostly random factors such as the direction the water enters from or the shape of the bowl itself (Joseph Castro, "Which Direction Does Toilet Water Swirl at the Equator?" livescience.com, October 25, 2011, accessed 1/20/2020).

7. Sharp-eyed readers will notice that this is close to the situation imagined by Galileo in 1632 in his *Dialogue on the Great World Systems*. Galileo claimed that a stone dropped from the top of the mast of a moving ship would strike the deck directly below the point from which it had been released. He reasoned that since the stone moved forward along with the ship, both stone and ship had the same horizontal, or *linear* velocity. But this is clearly wrong for an object moving radially within a rotating (noninertial) frame of reference, where the deck of a ship has a lesser (if negligibly so) *tangential* velocity than a man holding a stone at the top of the mast some 150 feet above.

8. Learning to soar like the birds, however, has not been without its perils. The first time a motorless, human-carrying aircraft took to the skies it killed its pilot, Willy Leusch. Leusch, a German ace pilot of World War I, was at the controls of a tailless glider launched from the top of a mountain in south-central Germany known as the Wasserkuppe, the highest peak in the Rhön mountains of that region. Taking advantage of the lift provided by winds blowing steadily upslope, Leusch successfully steered his experimental aircraft up to a height of several hundred feet. Flying straight into the wind, his ascent was as simple and swift as if he had been riding an escalator. But when Leusch attempted to execute a left turn—a relatively simple, banking maneuver that birds perform instinctively—his wings buckled under the differential, sideways air pressure on them, and he fell to his death.

Chapter 9

1. Inside the Earth's inner core there's also an "inner-inner core" whose crystalline structure is polarized in an east-west

direction as opposed to the north-south alignment of iron crystals in the outer-inner core (Xiaodong Song, "The Core Within Earth's Inner Core," Britannica. com, accessed 11/10/2020).

2. Newton identified red, orange, yellow, green, blue, indigo, and violet. Lately many people assert there are but six colors in the spectrum, omitting indigo. As the science writer Isaac Asimov once complained, "it has never seemed to me that indigo is worth the dignity of being considered a separate color. To my eyes it is merely deep blue" ("Indigo," en.m. wikipedia.org, accessed 12/4/2020). But Newton, influenced perhaps by music theory with seven notes on the major scale, not to mention the seven-days Creation of biblical tradition, saw—or thought he saw—seven.

3. *"[E]ine klassisch nicht bescreibbare Art von Zweideutigkeit,"* cited in "The Spinning Electron," *Nature Milestones*, Alison Wright, editor, nature.com, February 28, 2008, accessed 8/24/2020.

4. More than forty years later, Landé summed up the excitement of the early years of quantum physics: "[E]verybody was bothered by this anomaly [the apparent magnetic qualities of the electron] and had various wild ideas ... *and the wildest one certainly was that the electron was spinning around its own axis."*

5. A recent dilemma regarding subatomic spin was the "proton spin crisis" of the late 20th–early 21st centuries, solved, at least for the moment, by the supposition that an otherwise unaccountable portion of the spin of a proton is produced by the intrinsic angular momentum of its even smaller constituent particles, quarks and gluons.

6. Picture a spinning top. Grasp it with your right hand, thumb extended, fingers curled and pointing in the direction of rotation—to the right if it's spinning clockwise, to the left if counter-clockwise. Your thumb will then point either to "spin-up" or "spin-down."

7. The mass of an electron is $9.109 \times 10^{-[[31]]}$ kilograms, a number so vanishingly small that it is not considered in calculating the mass numbers of atoms.

8. In 2018, David Nadlinger of Oxford University captured a photograph of a single atom of strontium. Though Nadlinger's achievement has been rightly celebrated, he himself admits that the glowing, purple speck in the middle of the photo is not really the atom, not really the *Ding an sich*, only the light from surrounding lasers being re-emitted by it.

Works Cited

Chapter 1

Abrams, Joan. "Bingo Madness," *Lewiston Morning Tribune*, April 18, 1993, lmtribune.com, accessed 1/10/2021.

Conn, Philip. "Traditional Courtship and Marriage Customs in the Appalachian South," in Norbert Reidl, ed., "Glimpses of Southern Appalachian Folk Culture," appalachianhistory.net, accessed 11/23/2020.

"The Eco-Friendly Fly Deterrent," Hammacher Schlemmer Catalog, hammacher.com, accessed 11/23/2020.

Goto-Jones, Chris. "The Secret Life of Yo-Yos," *The Atlantic*, April 9, 2015, theatlantic.com, accessed 10/25/2020.

Heyd, Milly. "De Chirico: The Girl with the Hoop," *Psychoanalytic Perspectives on Art*, Vol. 3, edited by Mary M. Gedois, p. 97, books.google.com, accessed 7/1/2020.

Kastor, Elizabeth. "The Geppetto Mystique," *The Washington Post*, February 15, 1993, washingtonpost.com, accessed 1/17/2021.

Latson, Jennifer. "How Frisbees Got Off the Ground," *Time*, 1/23/2015, time.com, accessed 10/21/2020.

Lears, Jackson. "Fortune's Wheel," *Lapham's Quarterly*, laphamsquarterly.org, accessed 11/20/2020.

Lissaman, Peter, and Mont Hubbard. "Maximum Range of Flying Discs," *Science Direct/Procedia Engineering* 2 (2010), pdf.sciencedirectassets.com, accessed 10/22/2020.

Luscombe, Richard. "As Spinner Craze Goes Global, Its Inventor Struggles to Make Ends Meet," *The Guardian*, 5/5/2017, theguardian.com, accessed 11/1/2020.

Malone, Kenny. "Fidget Spinner Emerges as Must-Have Toy of the Year," Interview on "All Things Considered," npr.org, May 4, 2017, accessed 10/22/2020.

Mezrich, Ben. *Bringing Down the House.* New York: Simon & Schuster, 2014, books.google.com, accessed 3/13/2020.

Oliver, Valerie. "History of the Yo-Yo: Spinning Through the Ages," yoyomuseum.com, accessed 3/16/2019.

The Playground, vol. 16, 1916. Playground and Recreation Association of America, p. 231, archiv.org, accessed 2/22/2023.

Pogue, Trevor Keaton. "Inside the World of Competitive Yo-Yo," *Seattle Met*, February 28, 2018, seattlemet.com, accessed 11/1/2020.

Rosenberg, Anat. "Gyration Nation: The Weird Ancient History of the Dreidel," haaretz.com, December 14, 2014, accessed 3/10/2020.

Stevenson, Richard W. "Hula Hoop Is Coming Around Again," *The New York Times*, March 5, 1988, nytimes.com, accessed 7/2/2020.

Utton, Dominic. "Breaking the Bank at Monte Carlo," *Express*, August 2, 2018, express.co.uk, accessed 10/20/2020.

"Wellesley's 117th Annual Hoop Rolling," April 24, 2012, wellesley.edu., accessed 10/28/2020.

"The Yo-Yo Is on the Upswing," *The Los Angeles Times*, July 7, 1997, latimes.com, accessed 10/25/2020.

Chapter 2

Albury, Donald. "Braces: Part I," mathesontools.weekly.com, accessed 9/30/2020.

Berendson, Roy. popularmechanics.com, June 6, 2014, accessed 1/29/2019.

Bowden, Sue, and Avner Offner. "Household Appliances and the Use of Time: The United States and Britain Since the 1920s," *Economic History Review* XLVII (1994), jstor.org, accessed 9/12/2020.

Del Conte, Anna. *Anna Del Conte on Pasta.* London: Pavilion, 2016.

Gorelick, Leonard, and A. John Gwinnett. "Ancient Egyptian Stone Drilling: An Experimental Perspective on a Scholarly Disagreement," *Expedition Magazine* 25.3 (1983), penn.museum. org, accessed 9/20/2020.

Haga, Chuck. "It's Been a Tilting, Whirling Ride for 75 Years," *Minneapolis Star-Tribune*, August 31, 2001, web. archiv.org, accessed 2/28/20.

Hale, Tom. "New Footage Shows a Wild Orangutan Using a Saw to Cut a Tree in Half," *Spy in the Wild*, January 19, 2017, iflscience.com, accessed 7/19/2020.

Harvey, Katherine. *BBC History Magazine*, January 2020, historyextra.com, accessed 9/12/2020.

Hu, David. *How to Walk on Water and Climb Up Walls.* Princeton UP, 2018.

Knutson, Roger. *Flattened Fauna: A Field Guide to Common Animals of Roads, Streets, and Highways.* Berkeley, CA: Ten Speed Press, 1987.

Larkin, Jack. *The Reshaping of Everyday Life 1790–1840.* New York: Harper and Row, 1988, 162–3.

Maxwell, Lee. "Who Invented the Electric Washing Machine?" oldewash. com, accessed 9/10/2020.

Maynard, Micheline. "It's Time to Accept That the French Do Rolling Pins Better Than We Do," *The Takeout*, June 22, 2020, accessed 9/9/2020.

Miller, Stephen M. *Inspired Inventions: A Celebration of Shaker Ingenuity.* Lebanon, NH: University Press of New England, 2010.

O'Grady, Cathleen. "A Fidget Spinner to Detect Urinary Tract Infections," arstechnica.com, accessed 5/25/20.

Pappalardo, Joe. "The Secret History of Washing Machines," *Popular Mechanics*, July 26, 2019, popularmechanics. com, accessed 9/10/2020.

Peterson, Sarah. *The Cookbook That Changed the World.* Gloucestershire, UK: Tempus Publishing Group, 2006.

Powers, Keith. "Commonwealth Lyric Theater to Perform Traditional Ukrainian Opera in Newton, May 6, 2015," Newton.widkedlocal.com, accessed 9/15/2020.

Szondy, David. "Hard Grind: The Epic Journey of the World's Biggest Tunnel Boring Machine," *New Atlas*, May 1, 2017, newatlas.com, accessed 9/29/2020.

Chapter 3

Banu, Sabeeha. "20 of the World's Most Bizarre Competitions," *The Travel*, June 14, 2018, thetravel.com, accessed 3/21/2020.

Clarey, Christopher. "Appreciating Skating's Spins, the Art Behind the Sport," *The New York Times* 2/19/2014, accessed 10/11/2020.

Crittenden, Guy. "Reflections on Sufi Dance of Oneness at Menla," banafsheh.org, December 6, 2017, accessed 3/21/2020.

Defiyani, Eka, Porman Pangaribuan, and Denny Darlis. "Implementation of Raindrops Energy Collector Boards Using Piezoelectrictransducer," MATEC Web of Conferences, 2018, matec-conferences.org, accessed 10/12/2019.

Feuerlicht, Roberta Strauss. "Whirling Dervish; Still Mysterious and Exotic," *The New Times*, October 19, 1975, nytimes.com, accessed 3/23/2020.

Gooch, Brad. *Rumi's Secret: The Life of the Sufi Poet of Love.* New York: HarperCollins, 2017.

Heidorn, Keith C. "The Weather Doctor's Weather Almanac; Playing Through Snow and Ice: Part 2: Ice Skating and Sledding," islands.net.com, accessed 3/21/2020.

Kawar, Mary, and Sheila and Ron Frick. *Astronaut Training: A Sound Activated Vestibular-Visual Protocol for Moving, Looking and Listening.* Handbooks for Innovative Practice. Madison, WI: Vital Links, 2005.

Link, Jeff. "Playgrounds Spin into the Next Generation," goric.com, accessed 10/14/2020.

Macaulay, Alastair. "'Swan Lake' and Its 32 Fouettés," *The New York Times*, June 13, 2016, accessed 3/4/2020.

McClenon, Lee. "Understanding Energy Part 1: You Are as Powerful as a Lightbulb," sustainability.blogs.brynmawr.edu, September 31, 2012, accessed 4/1/2019.

Penn State Extension, "Better Kid Care: Spinning, Rolling, and Swinging! Oh My!" extension.psu.edu/programs/betterkidcare, accessed 3/9/2020.

Rieger, Andy. "Playground Memories of Many Americans," *Humanities*, Spring 2018, neh.gov, accessed 10/14/2020.

Ruh, Lucinda. *Frozen Teardrop*. New York: Selectbooks, 2011.

St. Fleur, Nicholas. "The Water Drop, It's the Greatest Dancer," *The New York Times*, March 26, 2019, nytimes.com, accessed 4/1/2019.

Shafqat, Mariam. "Sufism: Expressing Rumi's Thought by Painting Whirling Dervishes." *The Express Tribune*, July 30, 2020, tribune.com.pk, accessed 7/30/2020.

Strauss, Valerie. "Rethinking 'Ultra-Safe' Playgrounds: Why it's time to bring back thrill-provoking equipment for kids," *The Washington Post*, November 29, 2015, washingtonpost.com, accessed 9/3/2020.

Tabachnick, Cara. "Here's What You Need to Know Before Attending a Whirling Dervish Ceremony in Turkey." *Washington Post*, April 12, 2019, washingtonpost.com, accessed 3/22/2020.

Wallace, Arminta. "Giving the Dervishes a Whirl," *The Irish Times*, February 2, 2004, irishtimes.com, accessed 3/28/2020.

"A Whirling Dervish Puts Physicists in a Spin," *Sciencedaily.com*, November 26, 2013, accessed 3/3/2020.

Wylie, Robin. "Harnessing the Energy of Rain," eniday.com, accessed 4/1/2019.

Interchapter I

Mugglestone, Lynda. "Dictionaries: A Very Short Introduction," blog.oup.com, accessed 4/15/2020.

Robinson, David. "The Wheel of Fortune," *Classical Philology* 41, 4 (October 1946), jstor.org, accessed 4/16/2020.

Chapter 4

Adair, Robert. *The Physics of Baseball*. New York: HarperCollins, 1990.

Canales, Jimena. *A Tenth of a Second: A History*. Chicago: University of Chicago Press, 2009.

Harder, Ben. "Reinventing the (Fly) Wheel," *The Washington Post*, April 18, 2011, washingtonpost.com, accessed 12/11/2020.

Leary, Warren. "Physicists See Long Pass as Triumph of 3 Torques," *The New York Times*, January 2, 1996, nyt.com, accessed 1/3/2020.

Magnus, Heinrich Gustav. "On the Falling-Off Tendency of Rotating Bodies," *Annalen der Physik*, 1853, gallica.bnf.fr, accessed 12/22/2019.

Martin, Dan. "On the Origin and Significance of the Prayer Wheel," *The Journal of the Tibet Society*, 1987, repository.cam.ac.uk, accessed 1/1/2021.

Morrison, Faith. *An Introduction to Fluid Mechanics*. Cambridge: Cambridge University Press, 2013.

Prandtl, Ludwig. "Motion of Fluids with Very Little Viscosity," 1904, digital.library.unt.edu, accessed 11/28/2020.

Schaffer, Jay. "Coors Field: A Pitcher's Graveyard," researchgate.net, accessed 11/28/2020.

Tilden, William T. *Match Play and the Spin of the Ball*. American Lawn Tennis, 1925, books.google.com, accessed 11/29/2020.

Veilleux, Tom. "How Do Dimples in Golf Balls Affect Their Flight," *Scientific American*, September 19, 2005, scientificamerican.com, accessed 12/30/2019.

Vogel-Prandtl, Johanna. *Ludwig Prandtl: A Personal Biography Drawn from Memories and Correspondence*, trans. David Tigwell, Universitätsverlag Gottingen 2014, libnrary.oapen.org, accessed 112/23/2019.

Chapter 5

Anderson, Norman. *Ferris Wheels: An Illustrated History*. Bowling Green, OH: Bowling Green State University Popular Press, 1992.

Anthony, David. *The Horse, the Wheel, and Language*. Princeton: Princeton University Press, 2007.

Buck, Carl Darling. *A Dictionary of Selected Synonyms in the Principal Indo-European Languages*. Chicago: University of Chicago Press, 1949.

Czerski, Helen. *Storm in a Teacup: The Physics of Everyday Life*. New York: W. W. Norton, 2016.

Freud, Sigmund. *Moses and Monotheism*, trans. Katherine Jones. Leechworth: Hogarth Press, 1938. Universal Library, ia800500.us.archiv.org, accessed 1/8/2021.

Furukawa Electric Company News Release, April 15, 2015, Furukawa.co.jp, accessed 5/26/2020.

Köpp-Junk, Heidi. "Wagons and Carts and Their Significance in Ancient Egypt," *Journal of Ancient Egyptian Interconnections*, 2016, egyptian expedition.org, accessed 4/23/2020.

Larkin, Jack. *The Reshaping of Everyday Life 1790–1840*. New York: Harper and Row, 1988.

Peuchert, Will-Erich. *Schlesische Sagen Herausgegeben*. Munich: Diederichs Verlag, 1989.

Powell, Catherine. "The Nature and Use of Ancient Egyptian Potters' Wheels," amarnaproject.com, accessed 12/11/2020.

Purves, Dale, Joseph Paydarfar, and Timothy Andrews. "The Wagon Wheel Illusion in Movies and Reality," *Proceedings of the National Academy of Science USA* 93 (1996), wexler.free.fr, accessed 3/1/2023.

Rushton, W.A.H. "Effect of Humming on Vision," *Nature* 216 (1967), www.nature.com, accessed 6/9/2020.

Solnit, Rebecca. *The Faraway Nearby*. New York: Penguin Books, 2014.

Spierenberg, Peter. *The Spectacle of Suffering*. Cambridge: Cambridge University Press, 1984.

Vogel, Steven. *Why the Wheel Is Round*. Chicago: University of Chicago Press, 2016.

Watson, Lyall. *Heaven's Breath: A Natural History of the Wind*. New York: New York Review of Books, 1984.

Chapter 6

AccurateShooter.com. "Daily Bulletin," accurateshooter.com, June 3, 2008.

"Bullet Failure Midair Disintegration," posted by user woostri, September 17, 2010, Forum.snipershide.com, accessed 3/3/2021.

Collins, Graham. "Coriolis Effect," *Scientific American*, September 1, 2009, scientificamerican.com, accessed 2/24/2020.

Curtis, W.S. "Long Range Shooting: An Historical Perspective," archived in David E. Petzal, and Phil Bourjaily, *The Total Gun Manual* (Canadian Edition). San Francisco: Weldon Owen, 2014.

Greener, William. *The Gun and Its Development*. New York: Skyhorse (9th edition), 2013.

Jordan, John Richard, Jr. "Gatling, Richard Jordan," Dictionary of North Carolina Biography 1986, ncpedia.org, accessed 4/11/2020.

Kelly, Jack. *Gunpowder*. New York: Basic Books, 2004.

Love, H.D. *Vestiges of Old Madras* (1640–1800), Vol. 2 in the Digital Library of India, item 2015.70163, archive.org, accessed 2/25/2020.

Parramore, Thomas C. "Gatling Gun," ncpedia.com, accessed 4/12/2020.

Robins, Benjamin. "New Principles of Gunnery (1742)" in *Mathematical Tracts of the Late Benjamin Robins* (1761), internet archive, archive.org, accessed 11/30/2020.

Sagi, Guy. "The Forces at Work When a Bullet Leaves a Gun," *Shooting Sports USA*, July 3, 2017, ssusa.org, accessed 2/21/2020.

Stanage, Justin. "The Rifle-Musket vs. the Smoothbore Musket, a Comparison of the Two Types of Weapons Primarily at Short Ranges," scholarworks.iu.edu, accessed 2/29/2020.

Steele, Brett D. "Muskets and Pendulums: Benjamin Robins, Leonhard Euler, and the Ballistics Revolution," *Technology and Culture*, April 1994, accessed 4/5/2020.

Stephenson, E. Frank. "Gatling Gun," ncpedia.org, accessed 4/12/2020.

Turner, Warner Michael. "What Like a Bullet Can Undeceive?" english.yale.edu, accessed 11/30/2020.

Tyson, Neil deGrasse. Mobiletwitter. com, 8/11/2010.

Westwood, David. *Rifles: An Illustrated History of Their Impact.* Santa Barbara, CA: ABC-CLIO, 2005.

Interchapter II

Doppelbauer, M. "The Invention of the Electric Motor, 1800–1854," Karlsruhe Institute of Technology, eti.kit.edu, accessed 1/11/2019.

Harper, Jason. "The Dirty Dozen: Why 12-Cylinder Engines Remain Magical," *Robb Report*, September 26, 2017, accessed 2/17/2019.

Von Jacobi, Herman Moritz. "A Memorandum on the Application of Electromagnetism to the Movement of Machines," Potsdam, 1835, en.m.wiki source.org, accessed 1/31/2020.

Chapter 7

Alexander, David. *Nature's Flyers: Birds, Insects, and the Biomechanics of Flight.* Baltimore: Johns Hopkins University Press, 2004.

Danto, Arthur Coleman. "The Nation," 1996, Questia.com, accessed 4/19/ 2020.

Davis, Jennifer. "Pilots Love Aero Props," 2014, aero-motion.com, accessed 4/11/2019.

Didion, Joan. *The White Album.* New York: Farrar, Strauss, & Giroux, 1979.

Foumberg, Jason. "Iñigo Manglano-Ovalle," *Frieze Viewing Room*, November 17, 2013, frieze.com, accessed 4/28/2020.

Frazer, Sir James George. *The Golden Bough.* 1894, www.faculty.umb.edu, accessed 2/28/2023.

Hoppe, Jon. "Naval Aviation Oddity," *Naval History Magazine*, June 2021, www.usmni.org, accessed 2/26/2023.

McLean, Doug. *Understanding Aerodynamics: Arguing from the Real Physics.* Hoboken, NJ: Wiley, 2012.

Meloche, Monique. Meloche Gallery 2013, moniquemeloche.com, accessed 2/28/2020.

Regis, Ed. "No One Can Explain Why Planes Stay in the Air," *Scientific American*, February 1, 2020, scientific american.com, accessed 12/2/2020.

Shepherd, Dennis. "Historical Development of the Windmill," NASA, Lewis Research Center, 1990, osti.gov. accessed 6/17/2020.

Vanhoenacker, Mark. *Skyfaring.* New York: Vintage Books, 2015.

Vogel, Steve. "Foreword" in *Nature's Flyers: Birds, Insects, and Biomechanics of Flight*, by David E. Alexander, xiv. Baltimore: Johns Hopkins University Press, 2004.

Wright, Orville. "How We Made the First Flight," faas.gov, accessed 4/30/2020.

Chapter 8

"Biographies about the Eiffel Tower: Gaspard Coriolis," Wonders-of-the-world. net, accessed 12/3/2020.

Brown, Slater. *World of the Wind.* Indianapolis: Bobbs-Merrill, 1961.

Chantae. "Are the Sedona Vortexes in Arizona Real or Total BS?" chantae. com, April 22, 2020, accessed 11/8/ 2020.

de Coriolis, Gaspard-Gustave. *Mathematical Theory of Spin, Friction, and Collision in the Game of Billiards*, trans. David Nadler. San Francisco, 2005.

Donahue, Michelle. "Can We Capture Energy from a Hurricane?" *Smithsonian Magazine*, October 12, 2016, smithsonianmag.com, accessed 4/8/ 2019.

Garner, Dwight. "In Search of the Vortex Vibe in Sedona," *The New York Times*, April 9, 2006, nytimes.com, accessed 11/10/2020.

Kluger, Jeffrey. "A Bizarre 'Ice Circle' Is Turning Heads in Maine: Here's the Science Behind It," *Time*, January 16, 2019, time.com, accessed 1/16/2019.

Live Science Staff. "Secret Found to Flight of Helicopter Seeds," *Live-Science*, June 11, 2009, livescience.com, accessed 3/23/2019.

Niland, Lauren. "How the Mayan Calendar Was Brought to the World's Attention in 1987," *The Guardian*, December 20, 2012, theguardian.com, accessed 11/10/2020.

Sedonarocktours.com, accessed 3/29/ 2019.

Sharp, Andrew. "Gabby Street and the Dangerous Washington Monument Stunt!" baseballhistorycomesalive.com, accessed 2/26/21.

Trzcinski, Matthew. "Neil deGrasse Tyson on the Most Scientifically Inaccurate Movies Ever," *Showbiz Cheat Sheet*, May 28, 2020, showbizcheatsheet.com, accessed 2/28/2023.

Chapter 9

Barnett, Lincoln. "The Universe and Dr. Einstein," *Philosophymagazine*, September 4, 2018, riskservices.com, accessed 12/3/2020.

"Biologists Trace Evolution of Bacterial Flagellar Motors," *Sci News*, January 9, 2018, sci-news.com, accessed 10/4/2019.

Blakemore, Erin. "Christopher Columbus Never Set Out to Prove the Earth was Round," *History*, August 31, 2018, history.com, accessed 8/12/2020.

Born, Max. *Atomic Physics*. New York: Dover Publications, (1935) 1969.

Butterworth, John. *Atom Land*. New York: The Experiment LLC, 2018.

Choi, Charles Q. "Venus: The Hot, Hellish, Volcanic Planet," Space.com, accessed 8/11/2020.

Gamow, George. "Rotating Universe," *Nature*, October 19, 1946, nature.com, accessed 8/25/2020.

Gillen, Alan. *The Genesis of Germs*. Green Forest, AR: Master Books, 2007.

Giulini, Domenico. "Electron Spin or Classically Non-Describable Two-Valuedness," *Studies in History and Philosophy of Modern Physics* 39.3 (September 2008): 557–578. arxiv.org, accessed 8/23/2020.

Glatzmaier, Gary A. "The Geodynamo," websites.pmc.ucsc.edu, accessed 8/15/2020.

Goudsmit, S.A. "The Discovery of the Electron Spin." Translated J.H. van der Waals, lorenz.leidenuniv.nl, accessed 8/18/2020.

Kihara, Mary, et al. "Deletion Analysis of the Flagellar Switch Protein FliG of *Salmonella*." *Journal of Bacteriology* 182.11 (June 2000): 3022–28, accessed 1/12/2020.

LaBarbera, Michael. "Why the Wheels Won't Go." *The American Naturalist* 121 (March 1983): 395.

Macnab, Robert M. "The Bacterial Flagellum: Reversible Rotary Propeller and Type III Export Apparatus," *Journal of Bacteriology* 181.23 (December 1999): 7149–53, ncbi.nlm.nih.gov, accessed 11/5/2019.

Maddox, John. *What Remains to Be Discovered*. New York: Free Press, 1999.

Needham, Joseph, and Walter Pagel. *Background to Modern Science*. Cambridge: Cambridge University Press, 1938.

Panek, Richard. *The 4 Percent Universe*. Boston: Houghton Mifflin (Mariner Books), 2011.

Peebles, P.J.E. *Principles of Physical Cosmology*. Princeton: Princeton University Press, 1993.

Scoles, Sara. "How Vera Rubin Confirmed Dark Matter." *Astronomy*, October 4, 2016, astronomy.com, accessed 8/12/2020.

Tomonaga, Sin-itiro. *The Story of Spin*. Trans. Takeshi Oka. Chicago: University of Chicago Press, 1997.

Coda

Essen, Hanno, and Arne Nordmark. "A Simple Model for the Falling Cat Problem," *European Journal of Physics*, March 28, 2018, iopscience.iop.org, accessed 2/14/2020.

Index

Adair, Robert 79
air resistance 123, 136, 214
Alexander, David 147
Anderson, Norman 108
anemourion 158
Aristotle 2, 35, 217

backspin 89, 126, 214–215
ballet 4, 52, 56–59, 61, 63
Barlow's wheel 140, 142, 208
baseball 1, 4–6, 70, 73, 75–76, 78, 81–85,
	176, 179, 189, 214–215
Berlin Painter 14
boundary layer 75, 78–79, 81–82, 89, 214
brace and bit 41–42

Catherine wheel 118, 216
Cato 98
centrifuge 30–32, 213
Chamillionaire 96
Coriolis force 6, 55, 134, 173–178, 180,
	217–218
cyclone 168, 172, 178, 217

dark matter 195
Dionysus 8, 11, 15
downwash 147, 217
drag 6, 75–76, 81, 84, 89, 123, 134–136,
	214, 217
dreidel 9
drop spindle 105
dust devil 4, 181

electromagnetic field 190
electromagnetism 48, 138–141, 166
electron 172, 202–204, 206–210, 219
electron spin 206–209

fastball 73–74, 83–85, 214
Ferris wheel 54, 107–108
fidget spinner 18, 31, 99

flywheel 111–114
football 20–21, 88, 90–91, 133

Galileo Galilei 98
gaming wheel 26
Gamow, George 191
Ganymede 13–15
Gatling gun 131–33
Glatzmaier, Gary 190
golf ball 88–89
Gooch, Brad 51–52
Goto-Jones, Chris 16
gyroscope 1, 3, 6, 8, 60, 133, 150, 152–
	153, 213
gyroscopic compass 152

Hammacher-Schlemmer 22
Heyd, Milly 221
hoop 8, 11–15
hoop rolling 12–13
Hu, David 222
Hubble space telescope 127, 150, 172

ice disk 164

knuckleball 84, 90

lathe 34, 43–44, 187
Leonardo da Vinci 79, 98, 109, 119, 154
lift 20–21, 146–149, 183–184, 217–218
Ljubljana wheel 93, 96
Louis XIV 28

Macaulay, Alastair 59
Maddox, John 209
magnet 35, 138, 160–161, 205
magnetic field 35, 113, 138, 160, 190,
	205, 208
Magnus effect 82, 91, 126, 149
merry-go-round 61, 158, 203
Moskoestrom 164

Motor (electric) 35–36, 140–143, 160, 190, 197, 199–201
Moulin Rouge 157, 159
Mugglestone, Lynda 70

noria 108–109

Ovid 12, 45, 71, 106

Panek, Richard 193
panemones 159
peloton 69–70
photon 204–205
pirouette 56–59
Plato 8, 10, 29, 50, 106, 138, 187
playground 24, 55, 60–61
Pliny 36–37, 98
Prandtl, Ludwig 78–91
prayer wheel 115–117, 159
precession 9
propeller 144–148, 154–156, 200–201
Ptolemy 188, 195
Purves, Dale 101–102

quantum spin 207

record player 39, 47–49
rifle 3–4, 121–122, 126, 130–132, 135, 179, 215
rifling 119–122, 127
Robins, Benjamin 122–127
rolling pin 28, 32–34
rotational inertia 14–15, 21
roulette wheel 25–26
Ruh, Lucinda 63

salad spinner 29–32, 36, 38
samara 185–186
Sedona vortexes 166
skater 6, 62–64
skittles 21–22
slider 74, 82, 90

Solnit, Rebecca 107
Sperry Gyroscope Company 150
spin axis 8, 14, 56, 64, 82, 90
spinning wheel 70–71, 102–107
Sufi dancers 6, 52–54, 175

tennis ball 64, 75, 86–87, 125
The Third Man (film) 109
thrust 2, 147, 155
Tilden, William 86
tilt-a-whirl 32, 107
topspin 75, 82, 85–88
torque 9, 41, 57–58, 64–65, 91, 139, 143, 160–161, 197–198, 201, 212–213, 216–217
trade winds 179–180

Vanhoenacker, Mark 152–53
Virgil 8
Vogel, Steven 104
vortex street 181–182

wake turbulence 75, 81–82
washing machine 4, 36–38
Watson, Lyall 157
Wellesley College 14
Westwood, David 120–121
wet dog shake 30
wheel 23–27, 70–71, 92–110, 196, 215–216; see also Barlow's wheel; Catherine wheel; Ferris wheel; flywheel; roulette wheel
wheel bearing 4, 97–99
whirligig 7, 60, 109
whirling dervish 53; see also Sufi dancers
wind shear 169–170
Wright, Orville 147

yaw 91, 153
yo-yo 15–17